Pathways for Sustainable Sanitation

Achieving the Millennium Development Goals

Arno Rosemarin | Nelson Ekane | Ian Caldwell
Elisabeth Kvarnström | Jennifer McConville
Cecilia Ruben | Madeleine Fogde

EcoSanRes Programme | Stockholm Environment Institute | *Partner of* SuSanA

Publishing
London • New York

Published by **IWA Publishing**
Alliance House
12 Caxton Street
London SW1H 0QS, UK
Telephone: +44 (0)20 7654 5500
Fax: +44 (0)20 654 5555
Email: publications@iwap.co.uk
Web: www.iwapublishing.com

First published 2008

Layout: Ian Caldwell
Printed by Cambridge University Press, UK.

British Library Cataloguing in Publication Data
A CIP catalogue record for this book is available from the British Library

Library of Congress Cataloging- in-Publication Data
A catalog record for this book is available from the Library of Congress

ISBN: 1843391961
ISBN13: 9781843391968

Acknowledgements

This paper was generated as a visions document by the Stockholm Environment Institute within the Sustainable Sanitation Alliance (www.susana.org). The facts and data reported are from the cited published and electronic sources. The interpretations made in the document are those of the authors and do not necessarily reflect those of the partners of SuSanA. The report received valuable comments from within the group of seven authors and from the following external reviewers: **Rory Villaluna, Elisabeth von Meunch, Kirsten Doelle, Arne Panesar, Anna Norström, Pay Drechsel, Thor-Axel Stenström, Håkan Jönsson and David Crossweller.**

Stockholm Environment Institute
Kräftriket 2B
106 91 Stockholm
Sweden
Tel: +46 8 674 7070
Fax: +46 8 674 7020
E-mail: postmaster@sei.se
Web: www.sei.se and www.ecosanres.org

This publication should be cited as follows:
Rosemarin, A., Ekane, N., Caldwell, I., Kvarnström, E., McConville, J., Ruben, C. and Fogde, M. 2008. Pathways for Sustainable Sanitation - Achieving the Millennium Development Goals. SEI/IWA. 56p.

Cover photo by Nelson Ekane:
From Muea village near Buea, the provincial headquarters of the Southwest Province, Cameroon.

Table of Contents

List of Figures

List of Tables

List of Boxes

1. Introduction

1.1. Objectives of the paper

It is the objective of this paper to reflect on the progress being made in the sanitation sector and to examine what possible pathways exist to introduce sustainability in the United Nations Millennium Development Goal (MDG) process also showing that the target for sanitation ahs an impact on all the MDGs. The paper reviews the regional differences in some of the common health indicators such as diarrhoea mortality and DALYs in relation to inadequate hygiene and sanitation. It reviews the criteria surrounding the introduction of sustainable sanitation based in part on the work of the Sustainable Sanitation Alliance, what technical options exist in order to introduce more sustainable measures, what the costs can be compared to conventional approaches and what the potential impact of safe reuse of human excreta can have in agriculture, nutrition and food security. It reviews the WHO guidelines on safe reuse of human excreta and greywater in agriculture and also reviews a selection of sanitation planning tools. Finally it summarises what additional steps and pathways including financing and institutional arrangements that are needed in order to build in greater resilience and sustenance into the sanitation sector.

1.2. The MDG on water and sanitation

In 2000, the UN resolved to tackle the Millennium Development Goals or MDGs as they are commonly known. The MDGs are all-encompassing providing clear targets by 2015 to significantly reduce poverty, hunger, illiteracy, gender inequality, child mortality, disease and to ensure environmental sustainability and promote global partnership including open trade and financing. What the world didn't notice back in 2000 was that water supply and sanitation services had not been included as one of the environmental targets. And it was not until 2002 at the UN World Summit on Sustainable Development (WSSD) with the guidance of the WSSCC (Water Supply and Sanitation Collaborative Council) in Johannesburg that a target for these was adopted (MDG 7, Target 10). It turns out that sanitation is by far the largest of all the MDG targets affecting about 40% of the global population and possibly the biggest challenge since the sector has been so neglected and is riddled with development obstacles. It remains probably the largest killer of children and is still not a high priority among governments (WaterAid, 2008a).

World leaders came together in 2005 under the auspices of the 2005 World Summit at United Nations Headquarters to reiterate the MDGs as expressed in the Millennium Declaration and reaffirm their commitment to work towards achieving them, and also to assess the extent to which progress has been made since 2000. Likewise, on September 25, 2008 the UN High-Level Event on the MDGs took place to identify gaps and determine how to accelerate progress toward the MDGs. The Human Development Report (HDR) of 2006 (UNDP, 2006) provided a status report on the water and sanitation Target 10 of MDG 7, providing further necessary focus. These reviews have all concluded that the target for water supply is achievable but that for sanitation will not be met by 2015.

When it comes to the implementation of all the MDGs, the world seems to have forgotten the element of sustainability and what was resolved at the UNCED in Rio de Janeiro, Brazil, in 1992 and the WSSD in Johannesburg, South Africa, in 2002 (WSSD, 2002). Sustainability is absent from the MDG resolution of 2000 and the HDR of 2006. Neither does the JMP (Joint Monitoring Plan of WHO/UNICEF) include sustainability criteria in its definition of improved water supply and sanitation. The core of MDG 7 Target 10 is being tackled mainly using, for the most part, conventional approaches that lack technical innovation and social, economic and environmental sustainability (SEI, 2005). It is the objective of this paper to introduce such criteria and place it within a context of policy, planning and capacity development.

It is also high time for more comprehensive measurement tools in order to assess successes and failures within the sanitation sector. The global monitoring data on coverage have created a simplified division into "haves" and "have-nots" (Bartram, 2008). It is therefore a step in the right direction as is now the case in the latest JMP report of 2008 (WHO/UNICEF, 2008a) to monitor all forms of sanitation coverage, may it be sharing of communal latrines, unimproved, improved latrines or open defecation. The next step would be to provide some measures of installation longevity, functionality and sustainability.

Sanitation is a neglected and disjoint sector that in general is not properly understood nor prioritized by governments around the world. In addition it is not something that is well integrated into development programmes and tends not to be considered as a central concern for human well-being. Somehow it is "taken care of" in all sorts of ways from open defecation, to shared communal latrines, to water-borne household sewerage and centralized treatment systems. But in a way all of these are similar in that little or nor communication occurs between the stakeholders and the authorities. In many ways this is the "last chapter in human development" and there is much to be done (Rosemarin, 2007).

1.3. The International Year of Sanitation 2008

Perhaps because of the acknowledgement of the importance of sanitation and the staggering figure of those lacking basic sanitation, the UN declared 2008 the International Year of Sanitation to draw attention to the sanitation crisis and accelerate progress towards the sanitation target. Present global water and sanitation coverage figures clearly show that despite international commitment, enormous investment and concerted efforts to halve, by 2015, the proportion of people without sustainable access to safe drinking water and improved sanitation, there is still so much to be done to achieve timely success. The MDG sanitation target is an overarching one because of the importance of water and sanitation development as an instrument for sustainable development, economic growth, and poverty reduction. This makes MDG 7 crucial for the achievement of all the other MDGs (Table 1).

WHO/UNICEF (2008a) reported that in 2006, 884 million people in the world lacked access to improved water supply and 2.5 billion people lacked access to proper sanitation, with the great majority residing in the developing world. Only about one person in three in sub-Saharan Africa and South Asia has access to improved sanitation (UNDP, 2006). Furthermore, Black and Fawcett (2008) report that over 40% of people in the developing world still depend on a bucket, a bush, the banks of a stream, a back street or some other sheltered place for their several daily excretions. One in four people in Africa i.e. about 234 million people practice open defecation (WHO/UNICEF, 2008b). In India alone this number is over 600 million. Moreover, Clarke and King (2004) state that in much of the developing world, only a fraction of sewage and drainage water is treated before being discharged into waterways.

Table 1: MDG linkages

MDGs goals and targets	Linkages to environmental health/sanitation
Goal 1. Eradicate extreme poverty and hunger	A healthy environment means healthy people Ability to improve livelihoods Breaks the cycle of poverty/ill-health
Goal 2. Achieve universal primary education	Reduction in diarrhoeal and parasitic disease will result in increased attendance and participation in school School sanitation is an important determinant of girls' attendance
Goal 3. Promote gender equality and empower women	Environmental health risks reside disproportionately with women Effective interventions help to improve women's lives Empowers increased participation
Goals 4 and 5. Reduce child and maternal mortality	Appropriate environmental health interventions significantly reduce the deaths of children <5 yrs old related to unsafe water, sanitation and hygiene at the same time reducing the risk of maternity mortality.
Goal 6. Combat HIV/AIDS, malaria and other diseases	Preventive environmental health measures are as important as and at times more cost-effective than health treatments
Goal 7. Ensure environmental sustainability: *2015: Half the proportion (%) of people without access to safe drinking water and sustainable sanitation; 2020: A significant improvement in the lives of at least 100 million slum-dwellers*	Expressed in terms of environmental health improvements Environmental health measures such as sanitation contribute to the MDGs directly
Goal 8. Develop a global partnership for development	Strengthening cooperation between the UN and national governments in the sanitation area linking this to security, economic and social development, international law and human rights and democracy and gender issues.

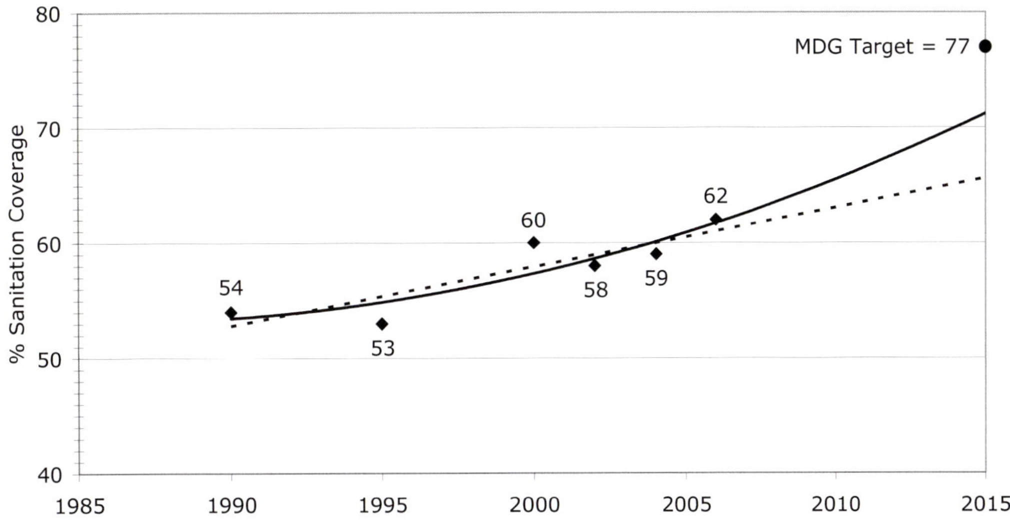

(Sources: 2000 data from JMP 2000 report, 2002 data from JMP 2004 report, 2004 data from 2006 JMP report, 1990 and 2006 data from 2008 JMP report, 1995 data from JMP website database accessed 7 August 2008).

Figure 1: Global sanitation coverage trends - linear (dotted line) and polynomial trend-lines

The magnitude of this ongoing sanitation crisis is so high in certain countries such as in India where Nadkarni (2002) report that 80% of the pollution load destroying the country's rivers is untreated human waste.

One of the most severe consequences of these inadequacies reported by WHO (2007) is that about 1.8 million children under five years of age die every year, or about 4900 young lives are lost daily from easily preventable diseases, such as diarrhoea. This makes it even more difficult to work towards the MDG 4 Target 5 - i.e. the reduction by two-thirds, between 1990 and 2015 of the under five years of age mortality rate. In addition, lack of access to safe water and proper sanitation in developing countries makes other population segments especially girls and women, suffer from poor health, diminished productivity and missed opportunities for education (WHO, 2007). This has a combined negative effect of slowing progress towards meeting all the MDGs and exacerbating the poverty situation (Satterthwaite, 2003).

1.4. Monitoring MDG 7 Target 10

The Joint Monitoring Programme (JMP) of the WHO and UNICEF was initiated in 1990 in order to monitor water supply and sanitation coverage around the world. This resulted in development country reports in 1991, 1993 and 1996. In 2000 the *Global Water Supply and Sanitation Assessment Report* was produced and JMP has since then produced reports in 2004, 2005, 2006 and 2008.

The baseline year for the MDGs is 1990 and the sanitation target is set to reduce the proportion of people without improved sanitation by 50% by the year 2015. The latest JMP report of 2008 (WHO/UNICEF, 2008a) shows that proportionate coverage is only slowly increasing with an increase between 1990 and 2006 of 8% (Figure 1) and at current trends the number of people without improved sanitation will probably be about 2.4 billion in 2015. It is of interest to plot the global sanitation coverage data made available by the JMP and also provide both a polynomial and linear interpolation plus projections beyond the latest data series of 2006. The polynomial provides a more optimistic scenario where the global MDG target is reached by 2019, while the linear projection predicts 2037 as the target year. The implications of using a more optimistic model now with seven years remaining to the end of 2015 could be significant since as with most human systems, growth in sanitation coverage may well be an exponential phenomenon once a certain critical mass has been achieved.

It is also of interest to plot the baseline 1990 global sanitation coverage data that have been published thus far by the JMP reports (Figure 2). There is some concern that the baseline for total coverage has been lowered from the original report in 2000 from 55% to 49% and then has risen again in the latest report for 2008 to 54%. The world total for rural communities drops from 35 to 25 and the rises again to 36 in the latest report. It should be understood that partner countries' databases are being updated from year to year and it is assumed for the purposes of the MDG projections that the latest version of the baseline is the one that is most accurate. The JMP database should not be overly criticised, however, taking into account the enormous task that it is undertaking. On the contrary it deserves praise and increased support in order to accomplish the task of monitoring progress being made as we get closer to 2015.

It is worth noting that some progress towards the achievement of the MDG 7 Target 10 is being made even in those regions where the challenges are greatest (Figure 3). Eastern, Southern and Western Asia, Northern Africa and Latin America and the Caribbean have been on track to meet the MDG 7 Target 10 (UN, 2006). The largest gains have been made in South Asia, where access to improved sanitation facilities more than doubled from 17% in 1990 to 37% in 2004, and in East Asia/Pacific, where, it rose from 30% to 51%. The improvements made in these regions were mainly driven by gains in India and China. For instance in India, sanitation coverage more than doubled – from 14% in 1990 to 33% in 2004, while in China sanitation coverage rose from 23% to 44% over the same period. Despite these increases in sanitation coverage in India and China, the majority of the people in these countries are still without access (UNICEF, 2006). All other developing regions have made insufficient progress towards this target. West/Central Africa, Eastern/Southern Africa and CEE/CIS (Central and Eastern Europe/ Commonwealth Independent States) are not on track to meet the MDG sanitation target. The least progress was made in CEE/CIS where coverage froze at 84%, and in Eastern/Southern Africa – where access improved only slightly, from 35% in 1990 to 38% in 2004, and where with population growth, the absolute number of people without sanitation increased by a third over the same period (UNICEF, 2006).

UNDP (2006) stresses that improved sanitation brings advantages for public health, livelihood and dignity – advantages that extend beyond households to entire communities. The need to provide sanitation to ensure safe drinking water and hygiene remains a huge challenge today in developing countries where often half the population lacks basic sanitation. Now, at the midway point between the year 2000 and 2015, halving the proportion of people without sustainable access to sanitation remains one of the MDG targets that is most off-track, with the crisis focused on sub-Saharan Africa, South Asia, and Oceania with 31%, 33%, and 52% sanitation coverage respectively in 2006. Sanitation coverage needed in these three regions to put them on track in 2006 was as follows: 50% for sub-Saharan Africa, 46% for South Asia, and 69% for Oceania. The ultimate MDG target coverage in 2015 is expected to be 63% in sub-Saharan Africa, 61% in South Asia, and 76% in Oceania. (WHO and UNICEF, 2008a).

The obstacles to accelerating the rate of progress in sub-Saharan Africa include conflict and political instability, high rates of population growth, and low priority given to water and sanitation (WHO and UNICEF, 2004). It should be added that the JMP coverage data is based on their current definition of improved sanitation which does not take into account safe discharge. If a more rigorous definition were to be used excluding installations that discharge untreated into drinking water sources, the coverage status would be much more off track for many other regions as well.

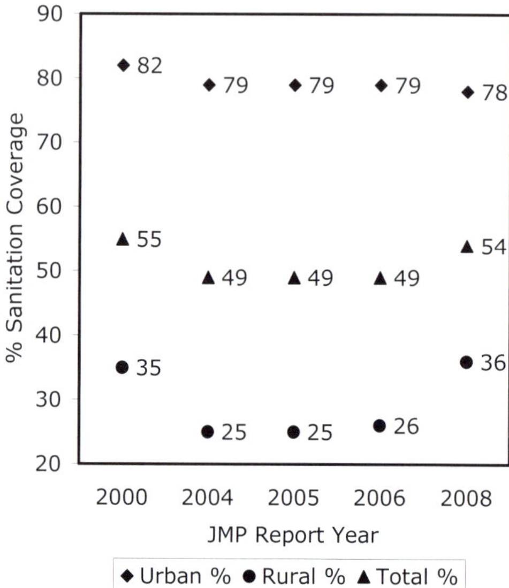

(Sources: GWSSA 2000 and WHO/UNICEF 2004, 2005, 2006 and 2008a)

Figure 2: Comparison of 1990 global sanitation coverage estimates

1.5. The gap between water supply and sanitation

Sanitation provision lags far behind access to water supply. Undoubtedly, this is a plausible reason why access to improved sanitation is not listed by the UN as one of the measures of progress towards the MDGs but is instead considered as one of the key challenges that has to be addressed in order to meet Target 10 of MDG 7 because of the astounding figure of those without access to basic sanitation. That sanitation was not included in the original MDG plan in 2000 does not make matters easier. Black and Fawcett (2008) reiterate that sanitation is invariably a poor relation to water in the public health engineering portfolio and receives far less resources.

Water supplies, for which there is strong demand and considerable political enthusiasm, consistently improve. Meanwhile, personal hygiene and waste disposal facilities, for which there is less overt and forceful demand, and substantially less political and donor interest, consistently languish. The gap between water and sanitation coverage ranges from 29% in East Asia to 18% in sub-Saharan Africa. In South Asia, access to improved sanitation is less than half that of water. These gaps matter not just because access to water and to sanitation is intrinsically important, but also because the benefits of improved access to water and to sanitation are mutually reinforcing. Thus countries that allow sanitation coverage to lag are destined to see the benefits of progress in water diminished as a result (UNDP, 2006).

Globally, 420 million people need improved water access and 1.052 billion people need improved sanitation systems between the years 2005 and 2015 to meet the MDG (Hutton and Bartram, 2008). The MDG 7 Target 10 will be missed by 600 million people if trends since 1990 continue (UN, 2007). Therefore, to meet this target, it was required that around 440,000 new people be served with improved sanitation facilities per day from January 2001 to December 2015.

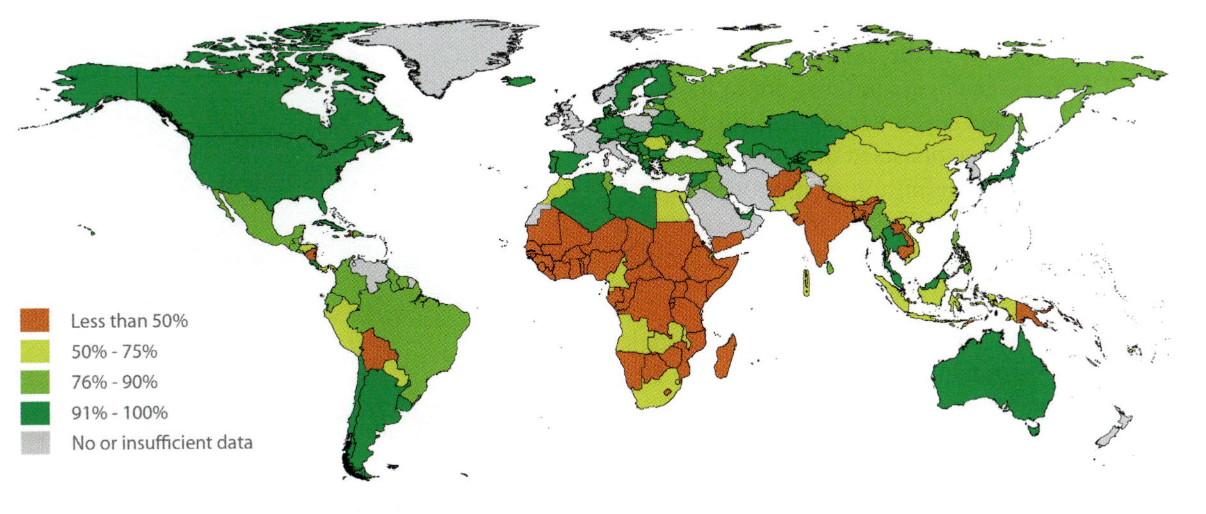

(WHO/UNICEF, 2008a)

Figure 3: Improved sanitation coverage in 2006

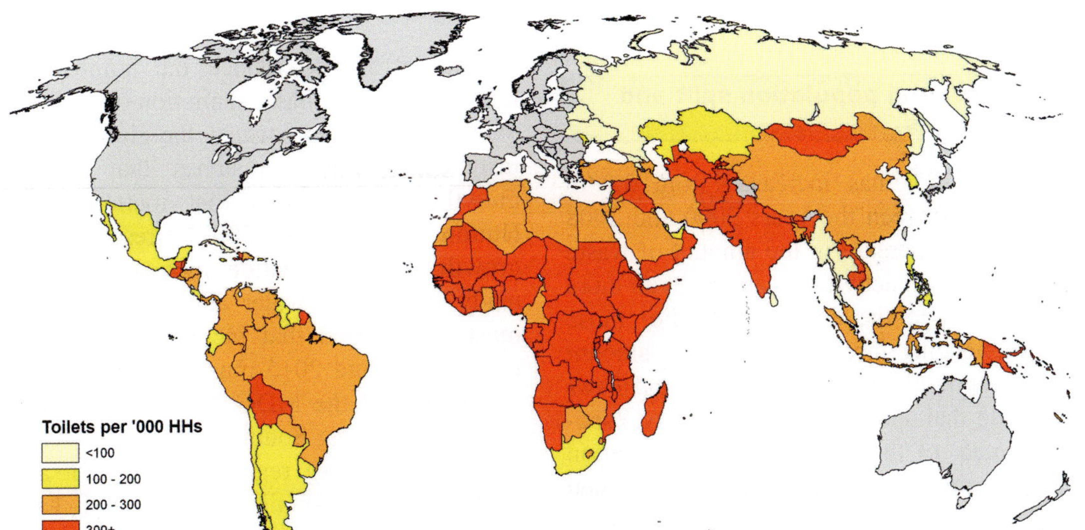

(Source: SEI, 2005)

Figure 4: Number of toilets per thousand households to be installed through 2015 to meet the MDGs

In order to meet the WHO and UNICEF target of "Sanitation for All" by 2025 an additional 480,000 people would have to receive improved sanitation facilities everyday from 2001 to 2025 (Mara *et al.*, 2007). According to WHO and UNICEF (2006), 1.6 billion more people need to gain access to improved sanitation over the period 2005 to 2015 to meet this target. This will reduce the population without adequate sanitation by 800 million i.e. from 2.6 billion in 2004 to 1.8 billion by 2015. But the world cannot cope with this gigantic challenge and the crisis continues. A measure of the MDG sanitation target can be expressed in terms of numbers of toilets needed per 1000 households (Figure 4). The greatest needs are in sub-Saharan Africa, South Asia, and particular countries in Latin America, West, East and Southeast Asia. Even if the polynomial projection for global increase in coverage turns out to be accurate, there will remain several countries still lagging. Again it should be reemphasised that these estimates are based on the JMP definition of improved sanitation and the backlog would be much higher if a more rigorous definition were to be imposed including sustainable sanitation criteria.

Between 1990 and 2006, the proportion of people without improved sanitation decreased by only 8%. Based on current trends, the total population without improved sanitation in 2015 will have decreased only slightly since 1990, to 2.4 billion. At this rate, the world will miss the MDG sanitation target by over 700 million people. Without an immediate acceleration in progress, the world will not achieve even half the MDG sanitation target by 2015. Thus to meet this target, at least 173 million people on average per year will need to begin using improved sanitation facilities (WHO/UNICEF, 2008a).

1.6. The urban/rural population split and growing slums

The world population has increased from 6.0 to about 6.7 billion between the years 2000 and 2008 with a concomitant increase in the amount of waste generated from human activities, causing even more pollution problems to the environment. For those countries experiencing increased GDP the amount produced per capita also increases. Black and Fawcett (2008) state that apart from the indignity, especially for women, of the lack of decent sanitation facilities, the world is daily becoming more crowded and more urbanized, with all the sanitary complications this represents. It is the urban areas of the developing countries where population will continue to increase (Figure 5) and this is where the most difficulties exist when it comes to sanitation.

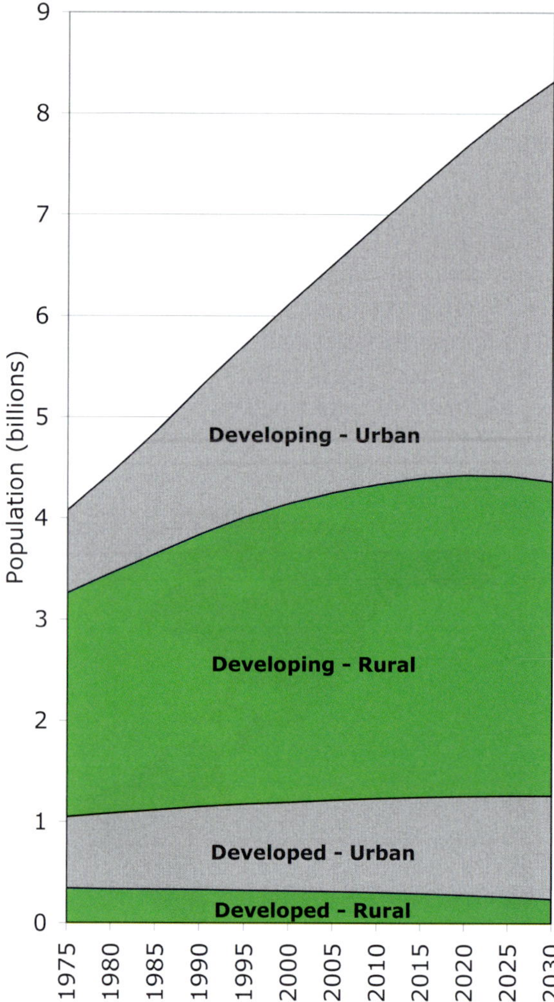

(Source: World Population Prospects: The 2006 Revision, medium variant)

Figure 5: Global urban and rural population trends in developing and developed countries.

Despite efforts to reduce the number of people without access to basic sanitation over the years, not much has changed because the global pace of toilet installations barely matches that of population growth in the places that matter. Furthermore, Hutton and Bartram (2007) stress that population growth and urbanization increase pressure on already inadequate water and sanitation services. It must be recalled that increases in population between 1990 and 2015 must also be included in the calculation of the MDG 7 Target 10. Sub-Saharan Africa (excluding South Africa) is the world's most rapidly urbanizing region (Tannerfeldt and Ljung, 2006), and almost all of this growth has been in slums, where residents face overcrowding, inadequate housing, and a lack of water and sanitation (UN, 2006). This means that more and more children will be raised in urban slums, without access to regular water supply, sanitation, health services, hygiene promotion and education. Over the next decade the population of developing countries is projected to grow by 830 million, with sub-Saharan Africa accounting for a quarter of the increase and South Asia for another third (UNDP, 2006). As for Asia, the level of urbanization in Africa is almost 40%, but with a much smaller urban population (0.3 billion).

According to WHO/UNICEF (2008a), 717 million additional rural inhabitants have gained access to safe drinking water since 1990. The estimated number of rural dwellers without improved drinking water supplies is 746 million, compared to 137 million urban dwellers. The world's urban population has increased by 956 million people since 1990 and during this period, 926 million urban inhabitants gained access to improved drinking water sources while the number of urban inhabitants without improved drinking water sources rose from 107 to 137 million predominantly in the developing world. The urban-rural disparities in the use of improved drinking water sources are highest in Latin America and sub-Saharan Africa. Oceania is the only developing region that has failed to lower use of unimproved drinking water sources since 1990 and thus half of the 9.2 million people in that region continue to use unimproved sources. While the world's urban sanitation coverage has increased to 79%, the rural coverage has risen much more slowly to 45%. One billion people in rural areas still practice open defecation which increases the risk of pathogen cross-contamination due to lack of containment. The increase in the number of urban dwellers using improved sanitation since 1990 (779 million people) has not kept pace with urban population growth of 956 million. Urban-rural disparities in the use of improved sanitation facilities are significant in most developing regions. Presently, the largest disparity between urban and rural sanitation coverage is found in Oceania, Latin America and the Caribbean, and Southern Asia. This disparity is smallest in East Asia.

From 2007 onwards more than half the world's human population, 3.3 billion people, was living in urban areas (UNFPA, 2007). The greatest challenge in water and sanitation provision will therefore be in the rapidly developing urban areas.

2. Linking Sanitation to Human Health

2.1. Impacts related to unsafe water, poor sanitation and poor hygiene

The common adage "Health is Wealth" is often used to mean good health is real wealth. In other words, this means that the money and time used in treating diseases could instead be used in carrying out productive activities. Wagstaff and Van Doorslaer (2003) report that in Vietnam health expenses are estimated to have pushed some 3 million people into poverty in 1993 and 2.7 million in 1998. According to Wagstaff and Claeson (2004), the international community's concern for health in the developing world is reflected in the MDGs in that nearly half the goals and targets concern directly or indirectly different aspects of health. In most developing countries, perhaps most easily discernible in the poorest countries, unclean water and poor sanitation exposes billions of men, women and children to a plethora of diseases that debilitate them and greatly reduce their productivity. These diseases even hasten the death of those with compromised immune systems such as the malnourished and HIV/AIDS patients.

Despite advances in science, engineering and legal frameworks, the majority of the wastewater from piped sewerage systems in the world is released into the environment without adequate treatment (Ujang and Henze, 2006). The same is true for most pit latrines and sludge management systems. Only a small percentage of global wastewater is treated using advanced sanitation facilities, mainly in developed countries. Faecal pathogens are transferred to the waterborne sewage system through flush toilets or latrines, and these may subsequently contaminate surface waters and groundwater (Prüss-Üstün and Corvalán, 2006).

Figure 6 illustrates the great divide between the developing and developed world in terms of public centralised sewerage systems. North America, East and West Europe, part of China and Japan dominate this development while most of Latin America, Africa, South and Southeast Asia lag behind. But as stated above, proper treatment of sewerage is a global problem with few exceptions.

As a result, the majority of the world's population is still exposed to waterborne diseases, and the quality of water resources has been rapidly degraded, particularly in poor developing countries (Ujang and Henze, 2006). UNDP (2006) describes unclean water as an immeasurably greater threat to human security than violent conflict. In addition to unclean water, lack of sanitation and poor hygiene are responsible for the transmission of diarrhoea, cholera, typhoid and several parasitic infections (UN, 2005).

Figure 6: Sewage sludge production from public sewerage systems mapped in terms of relative proportion of the global total for 1999

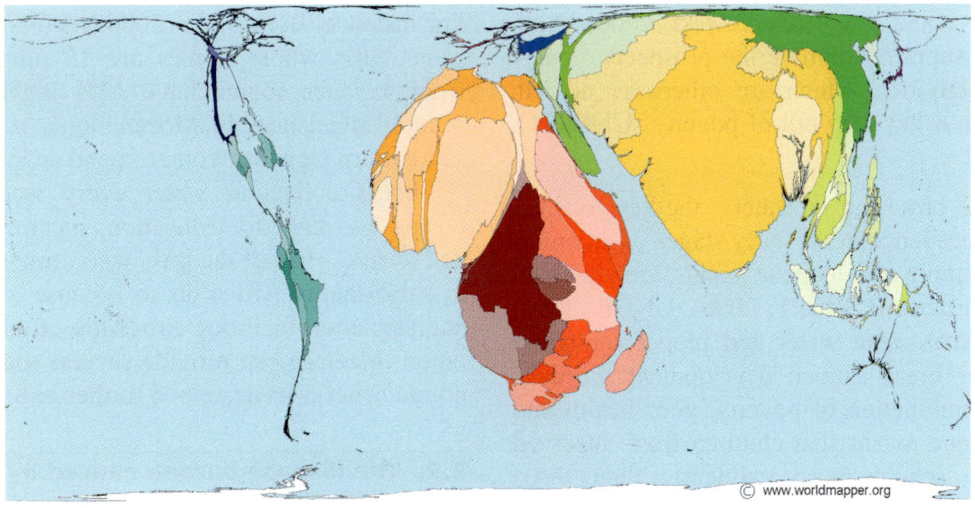

[Copyright 2006 SASI Group (University of Sheffield) and Mark Newman (University of Michigan)]

Figure 7: Diarrhoea-caused mortality mapped in terms of relative proportion of the global total for 2002

Moreover, WHO (1997) shows that the incidence of these diseases and others linked to poor hygiene and sanitation e.g. round worm, whip worm, guinea worm, and schistosomiasis is highest among the poor, especially school-aged children. These diseases have a strong negative impact on the health and nutrition of children and their learning capacities, and contribute to significant absences from school (Nokes and Bundy, 1993). Most significant, however is the persistence and wide distribution of diarrhoeal disease throughout the developing world.

Figure 7 illustrates again the great global divide in terms of diarrhoeal-caused mortality with Africa and South Asia showing great dominance. This correlates well with the fact that these two global regions are the ones most off-track in terms of meeting the MDG sanitation target.

Water-related diseases cost 443 million school days each year, and children in poor health suffer from reduced cognitive potential. This hurts their prospects for future earnings and makes continuing poverty more likely (Borkowski, 2006). About 3.5 billion people in the world are infected with helminth worm parasites, a fact that is not commonly realized (Chan, 1997; Capron *et al.*, 2004; Miguel and Kremer, 2004; UNESCO, 2006). These parasitic helminth infections often lead to severe consequences such as cognitive impairment, massive dysentery, anaemia and death of around 9400 people every year (WHO, 2004). In sub-Saharan Africa, schistosomiasis kills more than 200,000 people every year (Utzinger and Keiser, 2004). Infectious diarrhoea can be caused by bacteria (e.g. *E. coli*, shigellosis), viruses (e.g. norovirus, rotavirus), and protozoan parasites (e.g. amoebiasis, cryptosporidiosis, giardiasis) (OECD, 2007). Acute diarrhoea, as occurs with cholera, if left untreated can cause death within a day or less and has devastating impact on children (UNICEF, 2006). In 2005, West Africa suffered more than 63,000 cases of cholera, leading to 1000 deaths. Senegal was severely affected following rainy-season flooding in Dakar. During the first half of 2006, one of the worst epidemics to sweep sub-Saharan Africa in recent years claimed more than 400 lives in a month in Angola (UNDP, 2006).

Due to the interconnectedness between water, sanitation, health and poverty, lack of safe water supply and proper sanitation has much wider impacts than on just health alone (Table 2). Large-scale death and poor health are not only matters of concern in their own right but also act as a brake on economic development (Wagstaff and Claeson, 2004). The economic impact due to, for example, loss of tourism and export can be as dramatic as the cholera outbreak in Peru in 1990, which cost the national economy roughly US$1 billion in just 10 weeks – more than three times the total national investment in water supply and sanitation improvements during the 1980s. This was the first time in the 20th Century that epidemic cholera was identified in South America (Marshall, 2004; Bos *et al.*, 2005).

The interconnectedness as stated above, and the impacts that unsafe water and inadequate sanitation have on human health and general well-being makes it absolutely necessary to deal with all these issues or concerns together. Water quality and sanitation are irrevocably intertwined. Poor sanitation leads to water contamination. In many parts of the world, the main source of water contamination is due to sewage and human waste (Moe and Rheingans, 2006). Unsafe water and inadequate sanitation disproportionately impacts the poor in particular. About two-thirds of the people without access to a protected water source (piped water or a protected well) live on less than US$2 a day (Borkowski, 2006). Adequate water supplies improve the prospects of new livelihood activities, which are otherwise denied, and are often a key step out of poverty (UNESCO, 2006).

Although the provision of water - the key to life - may take precedence in the early stages of an emergency, sanitation and hygiene inputs are of equal critical importance (UNICEF, 2006). UNDP (2006) emphasises that clean water and proper sanitation can make or break human development. Furthermore, the combination of poverty, poor health and lack of hygiene means that children from unserved homes miss school more frequently than those whose families do benefit from improved drinking water and sanitation services. The resulting lack of education and social development further marginalizes the children and reduces their future chances of self improvement (WHO & UNICEF, 2005).

Table 2: Effects of poor sanitation and hygiene in Ethiopia

(Source: Ethiopian National Hygiene and Sanitation Strategy, 2005)

Health status	60% of the current disease burden in Ethiopia is attributable to poor sanitation where 15% of total deaths are from diarrhoea, mainly among the large population of children under five. Some 250,000 children die each year (WHO, 1997). High prevalence of worm infestations (causing anaemia) have a negative synergistic effect on the high levels of malnutrition.
Socio-economic cost	Negative synergy of diarrhoeal disease, malnutrition and opportunistic infections are known to have short-term health impacts and long-term debilitating effects. It is estimated that 4 in 10 children will not realise their educational potential which ultimately inhibits socio-economic development. Potentially productive time is lost to illness, caring for the sick and attending clinics. There are also financial costs of treatment i.e. for buying medicines and clinic attendance.
Poor educational performance	Worm infestations, anaemia and Vitamin A loss have been shown to decrease learning abilities among 4 in 10 girls. The lack of separate, private, secure, hygienic latrines, particularly in adolescence (during menstruation) is associated with a high drop out rate of girls. Girls end up staying at home to carry out household chores.

For instance, Borkowski (2006) reports that in Tanzania, girls whose homes are 15 minutes or less from a water source have 12% higher levels of school attendance. In Mozambique, rural Senegal, and eastern Uganda, women spend an average of 15-17 hours collecting water every week – which means less time for education, income generation, and leisure. Half of the girls who drop out of school in sub-Saharan Africa do so because of poor water and sanitation facilities. Provision of adequate sanitation therefore can provide several social and economic benefits as described further in Section 6.

2.2. The disease burden caused by lack of sanitation systems

It is of interest to see if it is possible using available data to assess if the reported increases in "improved" sanitation coverage have had any positive impact on human health and the environment knowing that there is a relationship between health status and sanitation coverage (Figure 8). An examination of mortality and DALYs databases related to diarrhoea provided by WHO shows that the latest data available for such an assessment is from 2002 (Prüss-Üstün *et al.*, 2008). So we are particularly ill-prepared to carry out an up-to-date critical assessment. Even trend analysis using historic data is difficult and has yet to be carried out. When compared to monitoring remediation progress for such diseases as malaria, turberculosis and HIV/AIDS, it is easily seen that sanitation-related diseases such as diarrhoea and helminth infections are not being given enough priority.

Current understanding of disease transmission indicates that disease burden is fully attributable to risk factors associated with water, sanitation and hygiene (Prüss-Üstün *et al.*, 2004). Every year millions of people, mostly in developing countries, die from diseases associated with lack of access to safe drinking water, basic sanitation and hygiene. Water-borne diseases in fact form the prime cause of premature death world-wide, especially for young children (WHO, 2002). Five times as many children die of diarrhoea as of HIV/AIDS (UNDP, 2006). Even though Mathers and Loncar (2006) project the risk of death for children younger than 5 years of age to fall by nearly 50% in the baseline scenario between 2002 and 2030, the under-five mortality rate remains a serious cause for concern because global progress is insufficient to achieve MDG 4. Many countries still have high levels of child mortality, particularly in sub-Saharan Africa (Figure 9) and South Asia, and in recent years have made little or no progress in reducing the number of child deaths (UNICEF, 2007).

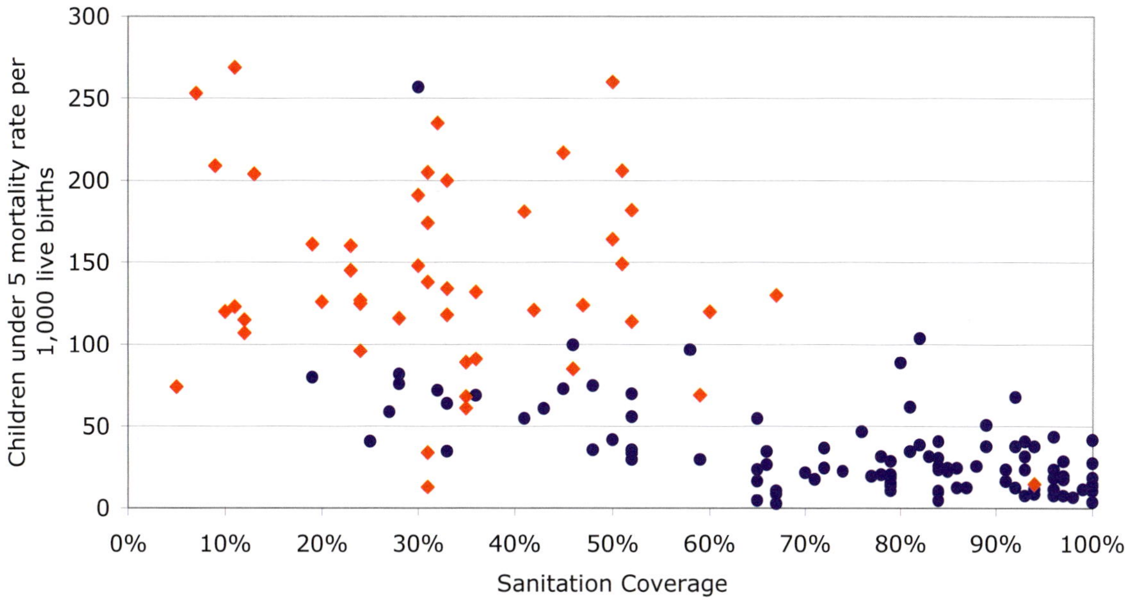

Each point is a single country with sub-Saharan Africa countries as red diamonds (Source data from WHO/UNICEF, 2008a and WHO, 2008)

Figure 8: Under 5 mortality compared to sanitation coverage for individual developing countries

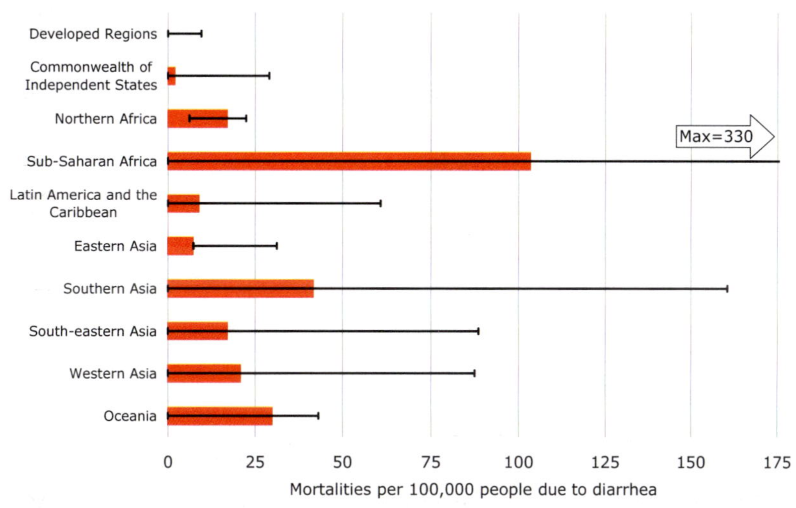

Line bars represent the range of individual country rates for each region (based on Prüss-Üstün et al., 2008)

Figure 9: Regional rates (red columns) of mortalities per 100,000 persons due to diarrhoea related to water, sanitation and hygiene for 2002

Not all diarrhoeal diseases are related to water. The share of diarrhoeal diseases attributed to water, sanitation and hygiene risk factors is likely to be approximately 60% in OECD (Organization for Economic Cooperation and Development) countries and around 88% at the global level. Diarrhoea, attributable to water and sanitation accounted for 5.3% of deaths and 3.5% of DALYs in European children aged 0-14 (Valent *et al.*, 2004). Each year, there are approximately 4 billion cases of diarrhoea worldwide. In Southeast Asia and Africa, diarrhoea is responsible for as much as 8.5% and 7.7% of all deaths respectively (WHO, 2008). The largest contributor to the burden of disease attributable to unsafe water, sanitation and hygiene is diarrhoeal diseases (OECD, 2007). This exceeds the disease burden of many major diseases, including malaria and tuberculosis, thus exacerbating the poverty situation in the poorest and most vulnerable regions of the world.

In the European and Central Asia region, the largest contribution to the burden of disease comes from those countries with low adult and low child mortality, with over 11,000 deaths and almost 500,000 DALYs. These countries are the WHO Eur-B subregion: Albania, Armenia, Azerbaijan, Bosnia and Herzegovina, Bulgaria, Georgia, Kyrgyzstan, Poland, Romania, Serbia, Montenegro, Slovakia, Tajikistan, The Former Yugoslav Republic of Macedonia, Turkey, Turkmenistan and Uzbekistan. This suggests that high potential reduction in deaths and DALYs could be made by the development of infrastructure and better personal hygiene. For instance, giving the entire child population in the Eur-B countries access to a regulated water supply and full sanitation coverage, with partial treatment of sewage, would save about 3,700 lives and 140,000 DALYs each year (WHO, 2004).

According to Prüss-Üstün *et al.* (2002), adding the disease burden from diarrhoea to other diseases related to water, sanitation and hygiene gives a total of 2.2 million deaths and 82.2 million DALYs per year i.e. 4.0% of all deaths and 5.7% of all DALYs. These estimates were based on 1999 disease statistics. WHO (2008) shows that the burden of disease linked to water and sanitation accounts for 58 million DALYs per year (94% of the diarrhoeal burden of disease) largely from unsafe water, sanitation and hygiene. Under-nutrition and malnutrition account for 2–3 times as much burden of disease as unsafe water and poor sanitation (Lopez *et al.*, 2006). Using more advanced methods, Prüss-Üstün *et al.* (2008) have included additional health effects, and taken into consideration the impacts of water, sanitation and hygiene on malnutrition and the consequences thereof in their latest estimates based on 2002 disease statistics. This gives a total of 98,789,000 DALYs per year for diseases related to water supply, sanitation and hygiene i.e. 6.6% of all DALYs, 2,412,000 deaths and 4.2% of all deaths. Water, sanitation and hygiene-related diarrhoea alone accounts for 52,460,000 DALYs and 1,523,000 deaths. Mathers and Loncar (2006) project a decrease in the burden of diarrhoeal diseases as years go by i.e. from about 64 million DALYs in 2002 to about 44 million DALYs in 2015 and about 29 million DALYs in 2030.

Sub-Saharan Africa remained the region with highest DALYs caused by diarrhoea related to lack of safe water, adequate sanitation and hygiene with approximately 34 DALYs per thousand people (Figure 10). This was followed by South Asia at 14, Oceania at 11, Western Asia at 7 and Southeast Asia and North Africa at 6. Here again the data show the extreme situation in sub-Saharan Africa. Unfortunately it is not yet possible to determine whether the overall increases in sanitation coverage are in fact lessening the disease burden. More data from recent years are required.

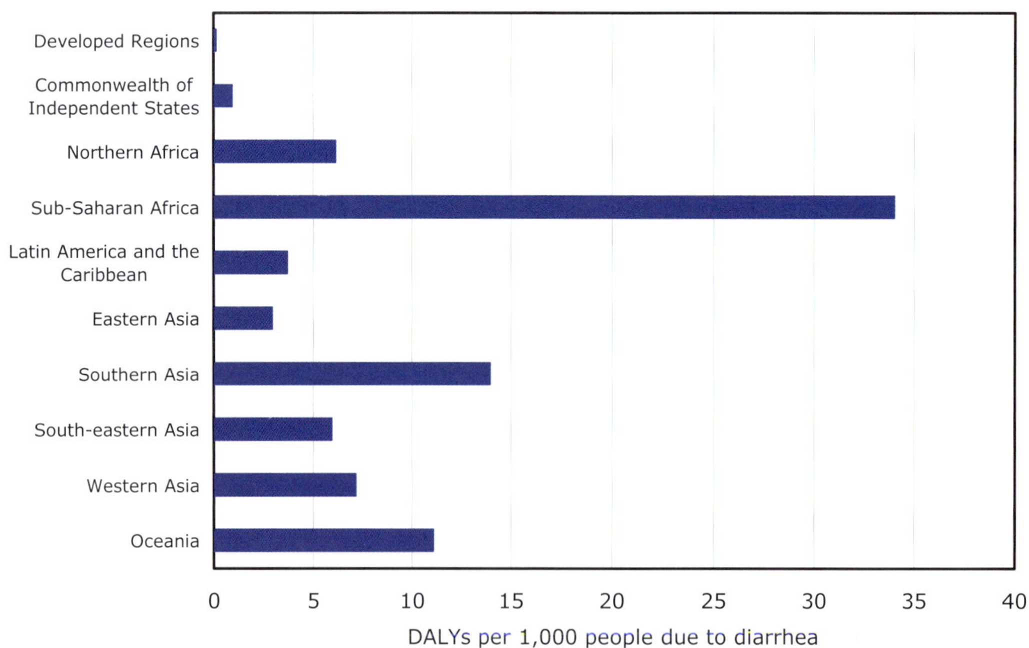

(based on Prüss-Üstün et al., 2008)

Figure 10: Regional DALYs per '000 persons due to diarrhoea related to water, sanitation and hygiene in 2002

3. Linking Sanitation to Food Security

A major driver for change within the sanitation sector is the potential use of excreta and wastewater for their nutrient content in the production of food. This is nothing new for many parts of the world in particular in East and Southeast Asia where human excreta have been used in agriculture for several centuries. Linking therefore global sanitation development to the global agriculture challenge is an important step to take in identifying drivers for sustainability.

3.1. The agriculture challenge

Currently some estimated 854 million people in 46 countries (Figure 11) are chronically hungry due to extreme poverty (von Braun, 2007; FAO, 2006), which is equivalent to around 15% of the world's population. About 2 billion people lack food security intermittently due to varying degrees of poverty (FAO, 2006). Each day 40,000 die of hunger and hunger-related diseases and famine threatens 9 African countries resulting in 20 million lives at risk.

At present, the earth is losing 25 billion tons of nutrient rich topsoil annually (WWI, 2005) which results in considerably reduced productivity, and hence decreased food security. According to UNEP (2007), globally some 2 billion hectares of vegetated land have been degraded since 1945, which is equivalent to 17% of the worldwide productively used land and 75-80% of Africa's farmland is degraded. Africa loses 30-60 kg of nutrients/ha/yr which is the highest rate in the world (World Bank, 2006). The highest rates of depletion are in Guinea, Congo, Angola, Rwanda, Burundi and Uganda at approximately 60 kg/ha/yr.

At present farmers worldwide use around 160 million tons of synthetically produced nutrients (N; P_2O_5; K_20) annually (IFA 2004; Figure 12), while at the same time conventional sanitation systems (piped systems and pit latrines) dump around 50 million tons of fertiliser equivalents with a market value of around $15 billion (Werner, 2004) into pits, and the surrounding environment including water bodies. In 2002/03 sub-Saharan Africa used 8 kg fertiliser/ha (World Bank, 2006) compared to South America (80 kg), North America (98 kg), Western Europe (175 kg), East Asia (202 kg), South Africa (61 kg) and North Africa (69 kg).

World fertiliser consumption is on the increase (Figure 12) and this will have further impacts on global prices. Fertiliser use has increased most significantly in the developing countries but since 1990 (hastened by the fall of the Soviet Union) has continued to decrease in most of the developed countries.

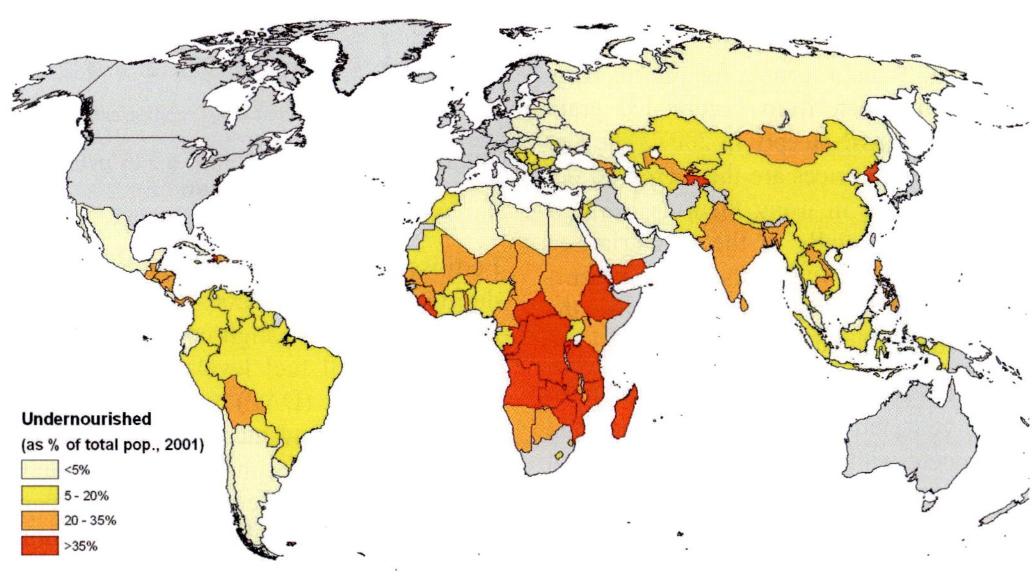

Shown as a percentage of total population as of 2001/2002 (Source: SEI, 2005)

Figure 11: Prevalence of undernourished in developing countries

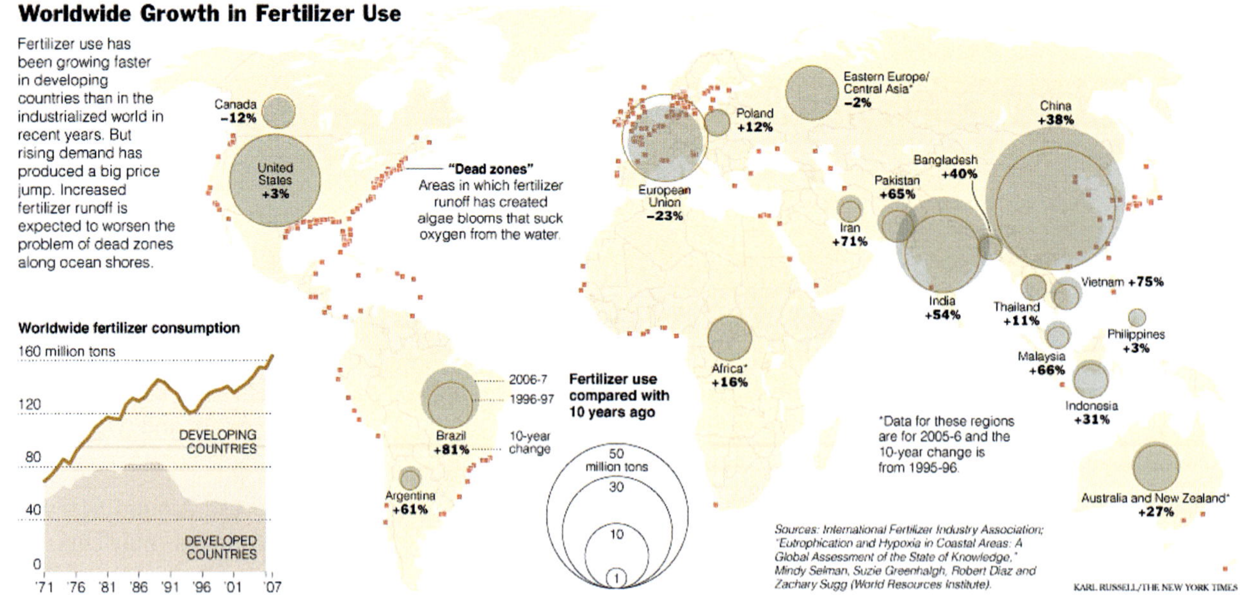

(Source: NY Times, May 2008)

Figure 12: World trends in fertiliser use

Over the past 10 years it has increased most in Asia where China and India dominate and has decreased most in Europe. World fertiliser consumption is expected to grow annually at about 1.7% from 2007/2008 to 2011/2012, equivalent to an increment of about 15 million tonnes. About 69% of this growth will take place in Asia and 19% in America (FAO, 2008a).

3.2. Rising food prices and the fertiliser market

The FAO Food Price Index rose about 80% during a 13 month period from January 2007 to February 2008 (FAO, 2008c) reflecting the general increases around the world. In particular it was grain prices that have been the main cause for these large changes. As can be seen from Figure 13, grain prices, especially rice rose steeply in 2007. The key reasons for higher grain prices are the increased demand and market value in using them as biofuels (von Braun, 2007) especially in the industrialised countries, the higher fossil fuel costs for their transportation, the soaring fertiliser prices and also the drop in value of the US dollar.

Rice is not directly affected by the interest in biofuels but is sensitive to fertiliser prices and maize and wheat also compete with rice as a basic source of dietary carbohydrate.

The global cost of chemical fertiliser (NPK) was USD 150/ton in 2006 and is now 3-5 times higher. In landlocked African countries it was USD 600/ton in 2006 due to poor transport infrastructure - rail & road and is now 4-5 times higher. Figure 14 illustrates the difference in fertiliser costs showing that Tanzania and Mali are paying 49% and 80% higher rates, respectively, than Thailand mainly due to additional transport costs and ancillary charges.

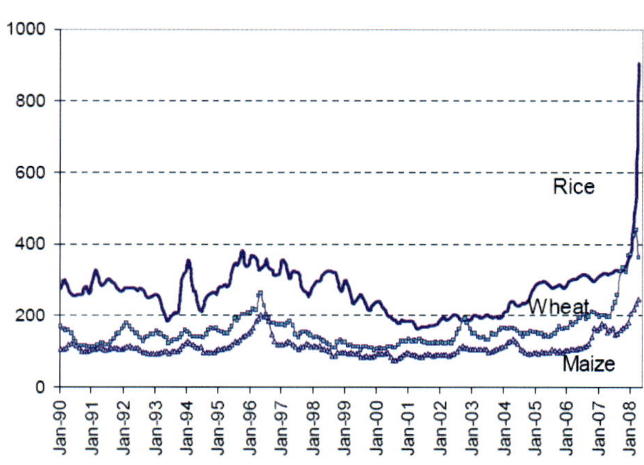

(Source: Brahmbhatt and Christiaensen, 2008. World Bank)

Figure 13: Trend in global grain prices in USD per ton

Fertiliser prices have been increasing significantly during the period Jan 2007 to April 2008 in part due to new markets for high-cost biofuels that are being priced at fossil fuel levels. The price of diammonium phosphate (DAP) has now exceeded US$ 1200 per ton, a five-fold increase since 2007 (Figure 15) (ICIS, 2008). Phosphate rock production in the United States has seen a marked decrease since 2006 with predicted depletion of economic reserves within 30 years (USGS, 2008; Figure 16). In addition, global supplies of sulphuric acid, used to produce phosphoric acid from the rock deposits, are scarce (Graff, 2007).

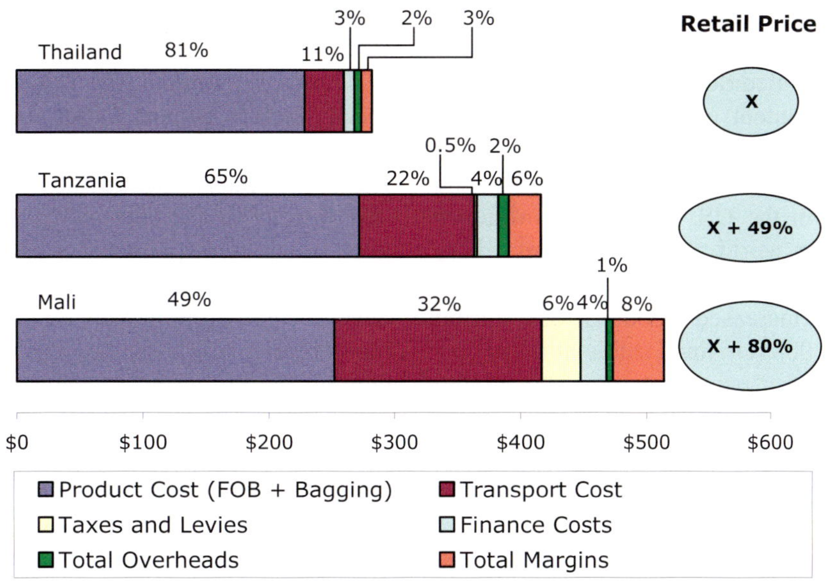

(Source: IFDC 2006)

Figure 14: Costs derived from transport, taxes, overheads, finance costs and margins cause fertiliser to cost much more in the poorest land-locked areas of the world such as Africa

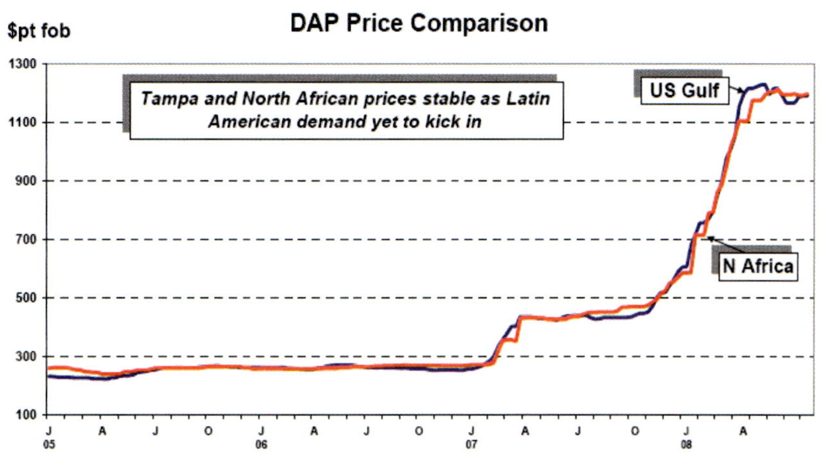

(USD Freight On Board per tonne. July 31, 2008. Source: ICIS)

Figure 15: Diammonium phosphate global bulk price trend

Natural gas, the main resource in the production of ammonia, has increased in price following oil and urea and ammonium rose about 4-fold during the period Jan 2007 to July 2008 (Figure 17 and Figure 18). These high prices have been further reinforced by the high grain and food prices in addition to continued increases in transportation costs.

3.3. The geopolitical perils of fertiliser availability and supply

Most of today's estimated global phosphorus reserves are being derived almost entirely from fossil deposits and it was estimated 5-10 years ago that these would last about 50 to 100 years (Steen 1998, Gumbo 2005). The phosphate reserves are mainly from fossil sediments. 89% of the global economic reserves are found in just 5 countries: Morocco/West Sahara, China, US, South Africa and Jordan (Figure 19). Economic reserves are defined by USGS as that part of the reserve base which could be economically extracted or produced at the time of determination. The term reserves need not signify that extraction facilities are in place and operative (USGS, 2006a). But at current rates of exploitation (increasing over 3% per year), the economic reserves will last no more than 50 years and the US economically viable reserves are expected to be depleted within 25 to 30 years (Rosemarin and Caldwell, 2007). That the US will become more and more dependent on imported phosphate is an explanation why in 2004 it signed with Morocco a free-trade agreement (Rosemarin, 2004). India is by far the largest and most vulnerable importer of phosphate having only limited domestic sources and it is almost entirely dependent on sources in Morocco. Phosphate supply will become more and more a central global geopolitical issue spurred on by the depleting reserves in the US. The strong influence of the US market on world phosphate prices where demand is greater than supply is seen from the fact that global prices increased about 5-fold during 2007 and have in 2008 remained stable at these levels. The prospects for potassium are comparable with phosphorus in that very few countries contain the abundant potash rock but the global reserves are estimated to last for the next 300 years (UNEP, 2002).

Global fertiliser **supply** is controlled mainly by just 9 countries and 70% of the world's phosphate rock is currently mined in just 4 countries: Morocco/Western Sahara, China, USA and Russia. Potassium supply is limited to 5 countries: Canada, Russia, Germany, Belarus and Brazil. 97% of nitrogen fertilisers are derived from methane (natural gas) which is available in significant quantities in greater than 60 countries which in theory should have a significant impact on competition. But the market price of NPK fertiliser is derived by the combined supply and demand of all three components acting together and it appears that competition at least on a regional basis is limited, contributing to the price increases shown above.

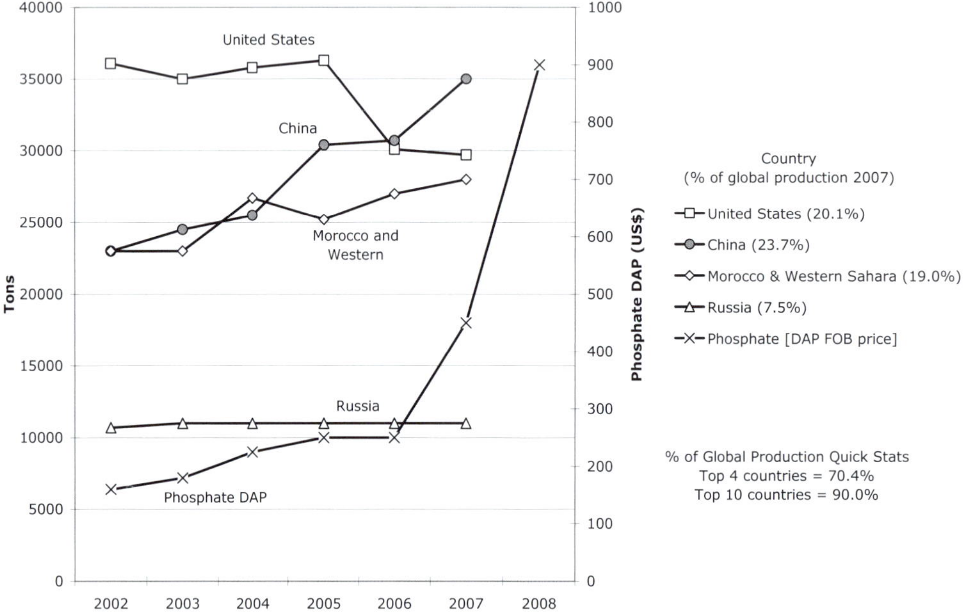

(Data sources: USGS [minerals.usgs.gov]; ICIS [www.icis.com])

Figure 16: Mine production of phosphate rock during 2002 and 2007 for the top 4 producers with the price trend for diammonium phosphate (DAP) in USD per ton between 2002 and 2008

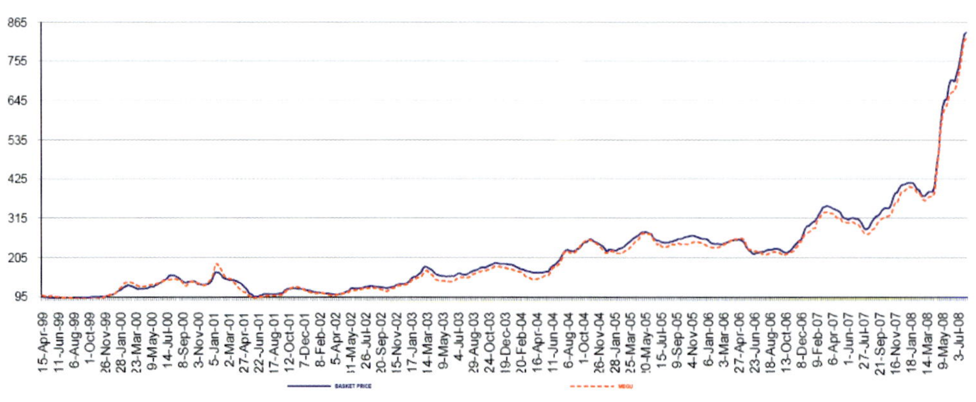

USD Freight On Board per tonne, as Basket Price and MEGU. August 1, 2008. (Source: IRM)

Figure 17: Granular urea US bulk price trend

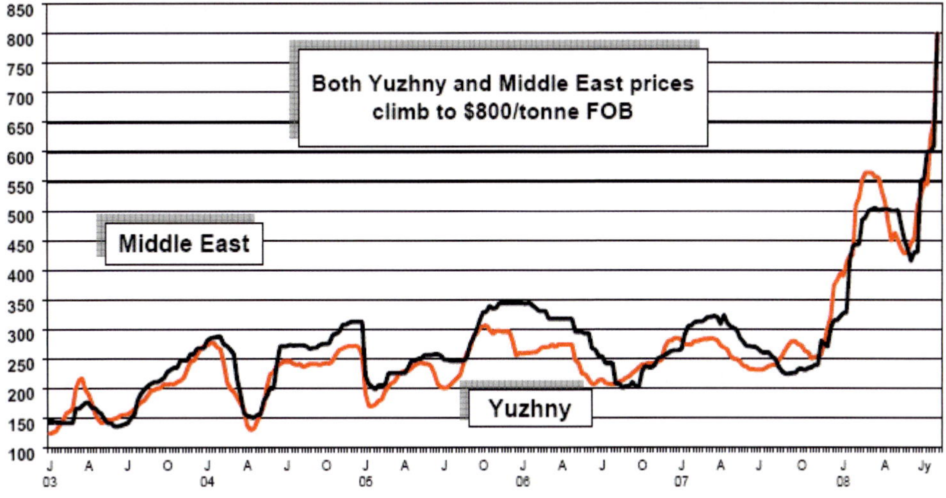

(USD Freight On Board per tonne. July 31, 2008. Source: ICIS)

Figure 18: Ammonia global bulk price trend

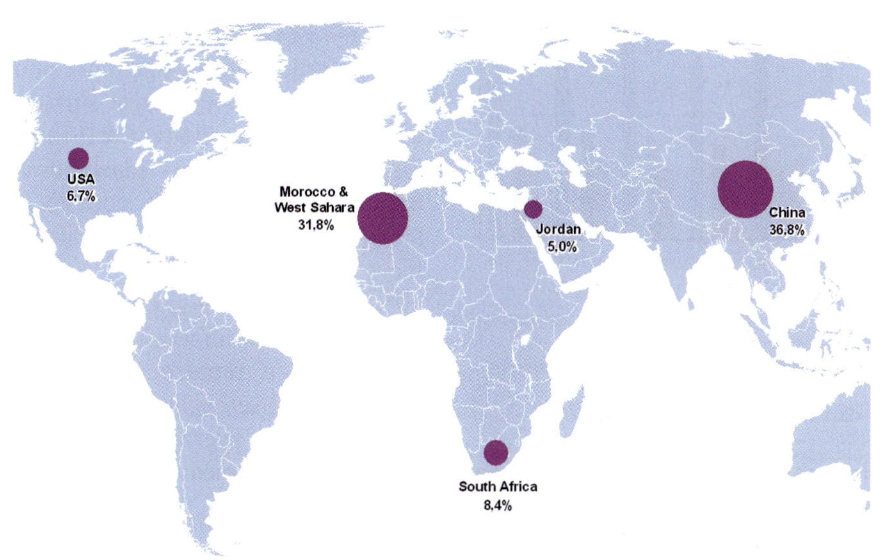

(Source data: USGS 2006b)

Figure 19: Geographic distribution of currently economic phosphate rock reserves

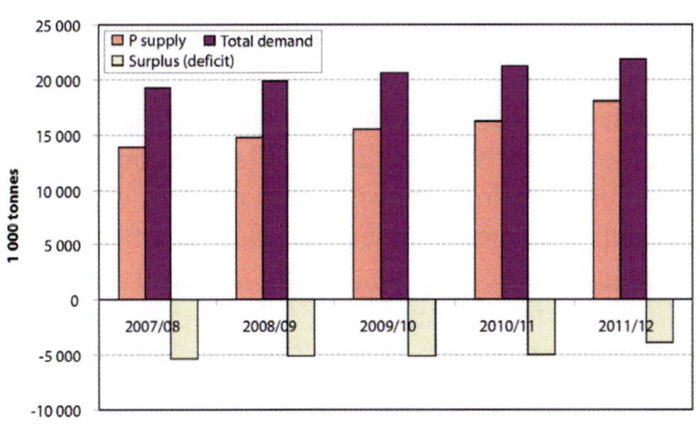

(Source: FAO, 2008b)

Figure 20: Current and projected supply and demand of phosphate fertiliser in Asia

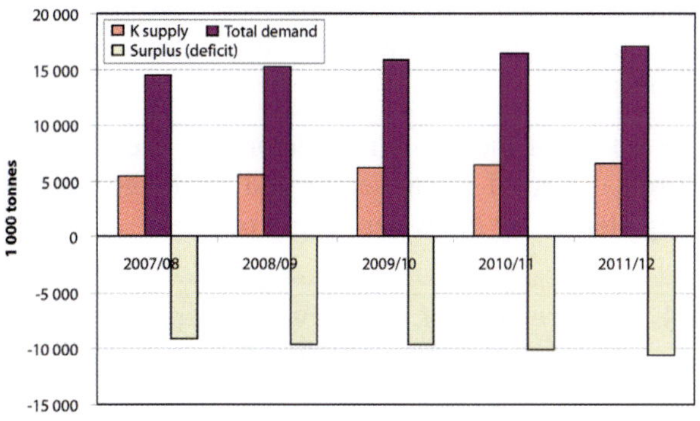

(Source: FAO, 2008b)

Figure 21: Current and projected supply and demand of potash fertiliser in Asia

The geopolitical constraints surrounding fertiliser become even more obvious when examining the significant supply deficit that Asia experiences for both phosphate and potash (Figure 20 and Figure 21). This is caused mainly by the fact that India lacks domestic or regional sources of phosphate and potassium and the same is true for potassium in China.

World phosphate fertilizer supply is forecasted to increase by 3.2% per year between 2007/8 and 2011/2012 while demand is predicted to increase at an annual rate of 2.4% (FAO, 2008a). This should help to stabilise world prices but phosphate is much more expensive per unit weight than nitrogen or potassium and has had the ability of lifting the global costs of NPK prepared products (Figure 22).

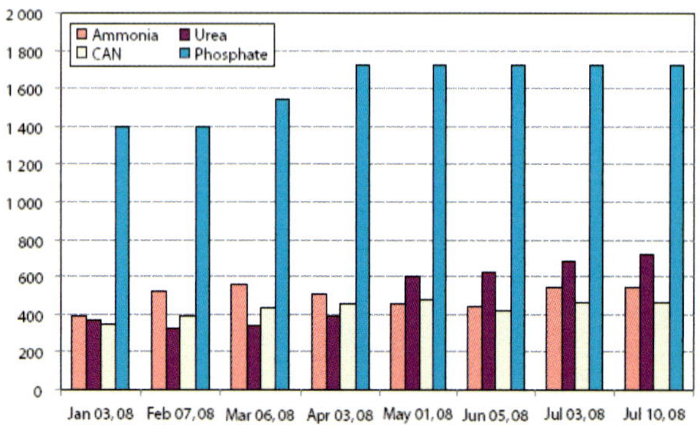

(Source: Yara [www.yara.com])

Figure 22: Global prices of fertiliser components (USD/ton)

Control of phosphate promises to be one of the key global resource questions which will have much larger social and economic impact than the present situation surrounding fossil fuels. As fertiliser and food become more and more expensive, countries will have to seek much more sustainable practices to optimise use of fertilisers in agriculture, to introduce sustainable agricultural practices, to promote large-scale composting from organic solid waste and to introduce large-scale nutrient reuse within the sanitation sector.

The future of the sanitation sector may therefore lie within the solid waste sector and no longer something steered by the water supply utilities.

3.4. Future role of conservation agriculture and productive sanitation

Each day, humans excrete in the order of 30g of carbon (90g of organic matter), 10-12g of nitrogen, 2g of phosphorus and 3g of potassium (Strauss, 2000). Most of the organic matter is contained in the faeces, while most of the nitrogen (70-80%) and potassium are contained in urine. Phosphorus is equally distributed between urine and faeces. The fertilising equivalent of excreta is nearly sufficient for a person to grow its own food (Drangert 1998) assuming collection and storage methods to prevent nitrogen loss are used.

Over the last hundred years major shifts in the developed countries have occurred in diet, food supply and sanitation practices. Food is traded globally now and most of the nutrients in sanitation systems are no longer reused. This means that the developed countries are very much handicapped if they decide to close the loop on critical nutrients like phosphorus using current day systems. A review of the situation for the Swedish city of Linköping between 1870 and 2000 was carried out by Schmid Neset *et al.* (2008) showing that in 1870, nearly all phosphorus contained in the human diet came from local agriculture and was recycled back to local agriculture, with a per capita dietary intake of 1.2 gP/person/day. Centralized sewers were installed in the 1950s and almost the entire phosphorus excreted was emitted to the nearby surface waters. In 2000 after phosphorus removal was installed, only around 20% of human dietary phosphorus was returned to agriculture (nearly 80% goes to landfill in sewage sludge). Dietary phosphorus intake increased over the 130 years by >25% to 1.6 gP/person/day. But 40% of food consumed in Sweden today is imported. The authors estimate that P-recovery from sewage could replace around one quarter of the mineral fertiliser necessary to produce food needs for the current average Linköping diet, or a higher proportion for a meat-free diet.

Nutrients from sanitation systems are being used in a clandestine and high risk fashion knowing that approximately 700 million people in 50 countries eat food from crops irrigated with untreated or inadequately wastewater from sewage systems on a total areal surface of at least 20 million ha (Scott *et al.*, 2004). Extensive use of wastewater also exists in the production of fish for example in the Calcutta wetlands and in the production of duckweed as fish and duck feed in Taiwan, Thailand and Bangladesh as reviewed by Strauss (2000). If sanitation systems were designed from the start with reuse of water and nutrients as a prime objective, the risk of infection by pathogens could be greatly reduced and this valuable water and nutrients could be exploited with lower risk to human health. The new WHO guidelines on reuse (Section 4.3) provide solutions to help reduce risk to the consumer, but for this to occur requires large outreach and awareness-raising implementation programmes.

As the price of fertiliser remains high and unaffordable for large parts of the developing world, alternative sources such as reuse from sanitation and solid waste systems will become more and more economic. The economic value of urine and composted organic wastes and faeces from both livestock and humans will make these products more and more attractive alternatives. There will be pressure to develop these options. It is thus highly relevant to examine the potential for replacement of chemical fertilisers by using conservation agricultural methods and so-called productive sanitation that allows for reuse of excreta in agriculture – this is of particular importance for the world's ca. two billion smallholder farmers (Onumah *et al.*, 2007) that cultivate limited areas and that at present cannot afford chemical fertilisers. That sub-Saharan Africa uses such low levels of chemical fertiliser (less than 10 kg of NPK/ha/yr) provides an immediate opportunity for the use of sanitation-based reuse systems. It was calculated that sub-Saharan Africa could become self-sufficient in fertiliser supply if it were to transform its sanitation systems to allow for reuse (Figure 23).

There are, however, major stumbling blocks preventing widespread development in these directions due to general ignorance and cultural taboos and attitudes about human excreta. There is a serious lack of capacity in the world today to carry out large-scale productive sanitation with agriculture applications. Although WHO in 2006 produced revised guidelines for the safe agro-reuse of nutrients from human excreta wastewater and greywater, the adaptation process has been slow. National policies and regulations are lacking to help promote and main-stream these practices. Much work through extension services and training are required before we can make the leap to close the loop on nutrients to benefit mankind.

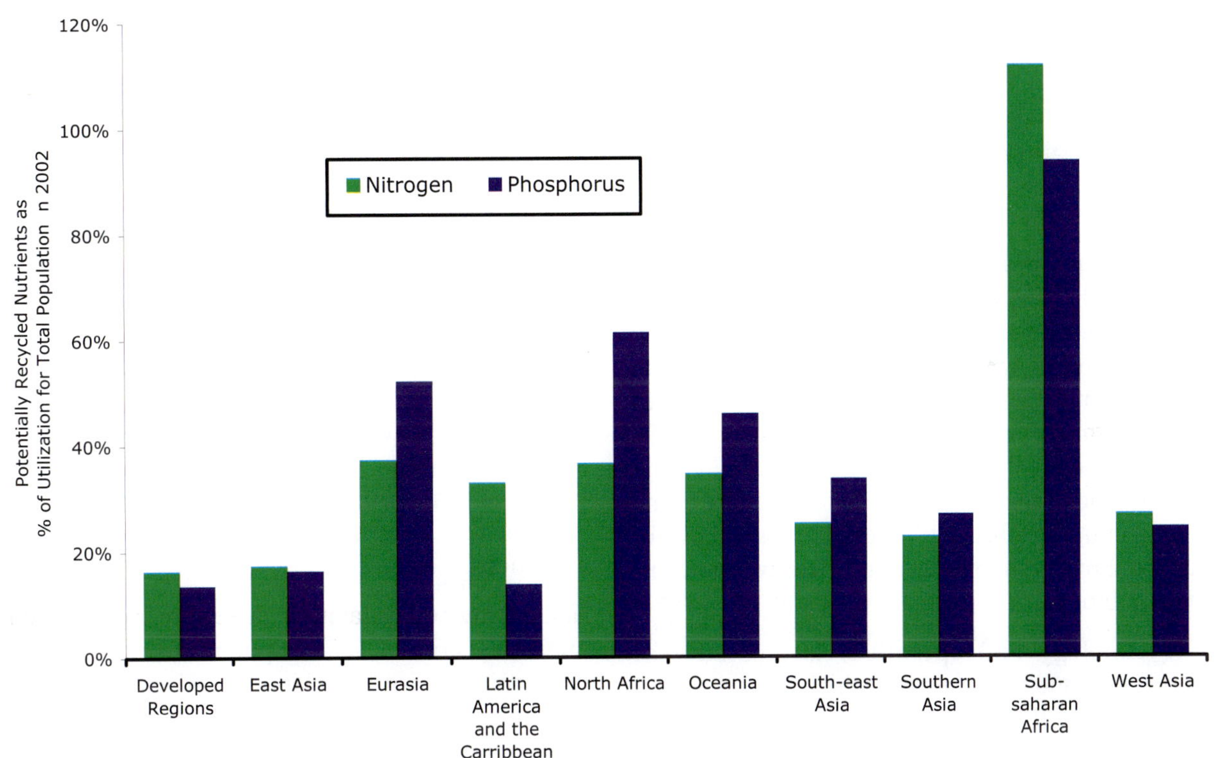

(Source: SEI, 2005)

Figure 23: Potential capacity of ecological sanitation systems to replace chemical fertiliser used in different world regions

4. Introducing Sustainable and Ecological Sanitation

4.1. Sustainable sanitation

"The main objective of a sanitation system is to protect and promote human health by providing a clean environment and breaking the cycle of disease. In order to be sustainable a sanitation system has to be not only economically viable, socially acceptable, and technically and institutionally appropriate, it should also protect the environment and the natural resources" - SuSanA (2008).

When improving an existing and/or designing a new sanitation system, sustainability criteria related to the following aspects should be considered:

(1) Health and hygiene: includes the risk of exposure to pathogens and hazardous substances that could affect public health at all points of the sanitation system from the toilet via the collection and treatment system to the point of reuse or disposal and downstream populations. This topic also covers aspects such as hygiene, nutrition and improvement of livelihood achieved by the application of a certain sanitation system, as well as downstream effects.

(2) Environment and natural resources: involves the required energy, water and other natural resources for construction, operation and maintenance of the system, as well as the potential emissions to the environment resulting from use. It also includes the degree of recycling and reuse practiced and the effects of these (e.g. reusing wastewater; returning nutrients and organic material to agriculture), and the protecting of other non-renewable resources, for example through the production of renewable energies (e.g. biogas).

(3) Technology and operation: incorporates the functionality and the ease with which the entire system including the collection, transport, treatment and reuse and/or final disposal can be constructed, operated and monitored by the local community and/or the technical teams of the local utilities. Furthermore, the robustness of the system, its vulnerability towards power cuts, water shortages, floods, etc. and the flexibility and adaptability of its technical elements to the existing infrastructure and to demographic and socio-economic developments are important aspects to be evaluated.

(4) Financial and economic issues: relate to the capacity of households and communities to pay for sanitation, including the construction, operation, maintenance and necessary reinvestments in the system. Besides the evaluation of these direct costs also direct benefits e.g. from recycled products (soil conditioner, fertiliser, energy and reclaimed water) and external costs and benefits have to be taken into account. Such external costs are e.g. environmental pollution and health hazards, while benefits include increased agricultural productivity and subsistence economy, employment creation, improved health and reduced environmental risks.

(5) Socio-cultural and institutional aspects: the criteria in this category evaluate the socio-cultural acceptance and appropriateness of the system, convenience, system perceptions, gender issues and impacts on human dignity, the contribution to food security, compliance with the legal framework and stable and efficient institutional settings.

Most sanitation systems have been designed with these aspects in mind, but in practice they are failing far too often because some of the criteria are not met. In fact, there is probably no system which is absolutely sustainable. The concept of sustainability is more of a direction rather than a stage to reach. Nevertheless, it is crucial, that sanitation systems are evaluated carefully with regard to all dimensions of sustainability. Since there is no one-for-all sanitation solution which fulfils the sustainability criteria in different circumstances to the same extent, this system evaluation will depend on the local framework and has to take into consideration existing environmental, technical, socio-cultural and economic conditions.

Taking into consideration the entire range of sustainability criteria, it is important to observe some basic principles when planning and implementing a sanitation system. These were already developed some years ago by a group of experts and were endorsed by the members of the Water Supply and Sanitation Collaborative Council as the "Bellagio Principles for Sustainable Sanitation" during its 5th Global Forum in November 2000:

(a) Human dignity, quality of life and environmental security at household level should be at the centre of any sanitation approach.

(b) In line with good governance principles, decision making should involve participation of all stakeholders, especially the consumers and providers of services.

(c) Waste should be considered a resource, and its management should be holistic and form part of integrated water resources, nutrient flow and waste management processes.

(d) The domain in which environmental sanitation problems are resolved should be kept to the minimum practicable size (household, neighbourhood, community, town, district, catchments, city).

What is truly sustainable in a given setting is to a large extent context-dependent and will vary widely between different settings. Consideration has to be given to the different issues related to health, environment, socio-culture, institutions, economy and technical function which are mentioned above as issues important to sustainable sanitation. For example institutional arrangements, responsibility division between users and authorities, the local climate, altitude, ground slope, soils, hydrogeology, housing density, liability to flooding, etc. will impact on what systems can be considered sustainable in any one setting. Moreover, local socio-cultural practices and preferences and socio-economic conditions, including willingness and ability to pay for the sanitation arrangement selected are other factors important to consider (Mara *et al.*, 2007). These factors will also vary, and thus impact what can be considered sustainable in a given setting.

Other more detailed criteria for what is sustainable sanitation exist in the literature (Bracken *et al.*, 2005; Winblad and Simpson-Hébert, 2004; UNESCO and GTZ, 2006). A detailed assessment of the elements required in gender equality within the water and sanitation sector, including empowerment of women is provided in the SEI report by Kjellén and Bernstein (2004).

4.2. Ecological sanitation

Ecological sanitation systems safely recycle excreta plant nutrients to crop production in such a way that the use of non-renewable resources is minimised. The statement "safe recycling" includes both hygienic and chemical aspects. Thus, the recycled human excreta shall be of high quality both concerning pathogens and heavy metals, etc. The statement "use of non-renewable resources is minimized" means that the gain in resources by recycling shall be larger than the cost of resources by recycling. An important boundary value for ecological sanitation systems is, thus, to be more resource-efficient, including crop production than sanitation systems NOT recycling human excreta, where equivalent amounts of nutrients are supplied using chemical fertilizers.

It can be noted that this definition of ecological sanitation takes into account two aspects of sustainable sanitation (health and environment) but not the other aspects (technical, institutional, economical or social). For true sustainability, ecological sanitation systems need to carefully address all these aspects.

Ecosan does not consist of a single toilet solution or a "one-size fits all" sanitation system. Instead, ecosan represents a wide range of options, appropriate for both poor and rich livelihoods, and rural and urban populations. In addition to protecting human health and the environment, it addresses a wide range of cultural needs such as indoor and outdoor installations, anal cleansing by using paper or water, and provides practical solutions to deal with odour arising from urine and faeces. Thus there is no single defined "ecosan toilet". One type of toilet with ecosan properties, where the pit is replaced by an above-ground vault is a good sanitation option especially for areas with shallow bedrock or high water tables where pit latrines cannot be installed (Moe and Rheingans, 2006). This sort of UDDT (urine-diverting dry toilet) has often been called "ecosan toilet". Other systems include soil-composting toilets like the Arborloo and Fossa alterna (Morgan, 2007), biogas fermentors attached to flush and pour-flush toilets and mixed water systems leading to constructed wetlands (Winblad and Simpson-Hébert, 2004).

The recycling of nutrients from urine and faeces is one of the key benefits of ecosan and this forms an interesting and important link between sanitation and agriculture. This is because the nitrogen, phosphorus and potassium found in urine is a valuable fertilizer and the high organic content of faeces makes the composted product – humus – an excellent soil conditioner. This has been recently called "productive sanitation" by proponents of sustainable agriculture whereby the various sanitation products (e.g. urine, faeces and greywater) are prepared for safe reuse. In addition, it is important to recover and reuse these nutrients to reduce the leaching of nutrients to surface and groundwater and to reduce the drain on natural reserves and lessen the dependency on chemical fertilizers. An average human produces 500 litres of urine and 50 litres of faeces per year. This is equivalent to about 5.5 kg of NPK (4 kg of nitrogen, 1 kg of potassium and 0.5 kg of phosphorus) per capita per year varying from region to region depending on food intake. Calculations show that sub-Saharan Africa could become self-sufficient in fertilizer supply if it were to adopt ecological sanitation practices (SEI, 2005).

For two billion smallholder farmers in the world, today's chemical fertilisers are not affordable. Alternatives using organic fertilisers from composting and productive sanitation may be the only options. Some countries and cultures have been recycling human excreta for agriculture for thousands of years, especially in China and Southeast Asia, but often excreta have not been properly sanitized therefore propagating disease. Well-planned and efficiently working ecosan solutions can provide a series of additional benefits to those using conventional approaches. These include:

- Recycling of nutrients derived from human excreta for local agriculture (especially important in regions with poor soil fertility and poverty);
- Permanent installations (conventional pit latrines last 5-10 years and then are often abandoned);
- Prevention of downstream ground and surface contamination by nutrients (e.g. nitrate), organics and pathogens (ecosan systems provide a high level of pathogen kill-off);
- Containment improvements over leaky septic tanks and sewage systems;
- Lower cost latrines in urban areas not requiring large-bore sewage pipe collectors and large treatment plants;
- Savings of domestic water (especially important in drought-prone regions);
- Biogas production for cooking, heating and lighting;
- Woody bioenergy from constructed wetlands;
- Alternatives to pit latrines in areas of high water tables and flooding

Thus, a functional ecosan system has the capacity to meet the WHO guidelines for safe reuse of human excreta, wastewater and greywater and plays an essential role in achieving the MDGs. Lopez *et al.* (2006) estimate a 3-4 times burden of disease with conventional sanitation solutions compared to those with better containment. Ecosan emphasises the aspect of source separation in order to allow for containment, sanitisation and reuse of excreta following treatment. In wet ecosan, urine, faeces and even greywater can be collected and treated together for example using constructed wetlands. The objective is to protect human health and the environment while reducing the use of water in sanitation systems and recycling nutrients to help reduce the need for artificial fertilizers in agriculture. Ecosan represents a conceptual shift in the relationship between people and the environment and it is built on the necessary link between people and soil. The conceptual model of ecological sanitation is shown in Figure 24. Ecosan is a closed-loop system, closing the nutrient and water cycles.

Interest in ecosan can be triggered by using community awareness participatory methods. According to Kar (2005), calculating the amount of faeces produced by open defecators in a community can help to illustrate the magnitude of the sanitation problem. Households can use their own methods and local measures for calculating how much they are adding to the problem. The sum of the households can be added up to produce a figure for the whole community. A daily figure can be multiplied to know how much faeces is produced per week, per month or per year. These quantities can add up to tonnes which may surprise the community.

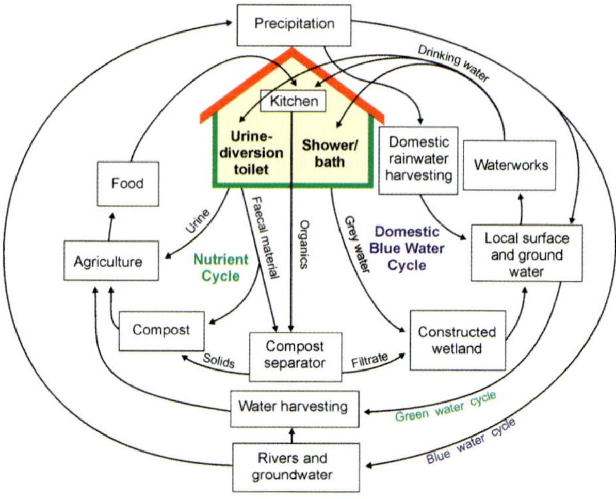

(expanded from Oldenbrg, M. [Otterwaser])

Figure 24: Complete household-based urine-diversion ecosan and eco-water use, closing the nutrient and water cycles

Lessons from the low-cost ecosan project in Malawi show that ecosan works well when it builds on existing practice; champions are important to ensure ongoing success – women make good champions; word of mouth and observation are effective communication strategies; people need time for experimentation; cheap, home produced fertilizer is a powerful selling point; ecosan boosts declining soil fertility and is more acceptable to farmers who experience this problem, cannot afford chemical fertilizers and recognize the ecosan potential; but ecological sanitation does not suit everybody (WSP, 2007b).

Figure 25 illustrates the effect of urine additions to maize growth in nutrient-poor dryland soils. Significant gains can be achieved by using urine instead of chemical fertiliser at the same sort of nitrogen rates. Urine of course contains much more than nitrogen and is a complete balanced fertiliser including phosphorus, potassium, sulphur and micronutrients and provides a long-term solution to soil fertility. Introducing urine-diverting sanitation systems would offer small-holder farmers an affordable source of fertiliser which would have significant impacts on livelihood development and nutrition.

Figure 26 illustrates the effect of applying approximately one day's urine from one person on maize growth. This amount of urine (1.75 L) applied over the growth period of 3.25 months resulted in almost 1 kg of cob fresh weight. Extending this to what can be gained from the urine from one person produced over an entire year provides enough to fertilise 300-400m^2 of crop.

Epworth, Harare: unfertilized maize (left-side) urine-fertilized maize (right-side) (Source: Morgan, Aquamor)

Figure 25: Maize crop 31 Jan 2005

Just over one day's urine (1750 mL) from one person resulted in almost one kilo additional fresh weight compared to the control receiving no urine (Source: Morgan, Aquamor)

Figure 26: Effect of addition of varying amounts of urine on the growth of one maize plant consisting of three cobs over a 3.25 month growth period

Figure 27 is a comparison of chemical nitrogen fertiliser use compared to potential nitrogen produced from sanitation systems (SEI, 2005). The 3000 kcal/day line is the ideal level of nutrition according to UN standards and the curved vertical line is the level where utilization and potential recycling is in balance. Countries closest to the crossing point of the two lines can be considered most sustainable. This divides the world into three categories – the undernourished and underfertilised to the bottom left, the undernourished and over-fertilised to the bottom right and the overnourished and overfertilised to the upper right. This shows, for example, that the developed countries to the top right are exporting food they produce and are producing large amounts of meat with the fertiliser much of which doesn't enter the human digestive system. The opposite extreme is the poorest countries that use very low amounts of fertiliser and are experiencing malnutrition. The category to the bottom right is an interesting one where large amounts of fertiliser are used and domestic requirements are still not met.

Hand-in-hand with the introduction of ecosan or productive sanitation systems is the need to promote resource-based agricultural practices instead of the more common commodity-based approaches, especially in developing countries. This would require the following policy shifts:

- education and encouragement of farmers to adopt resource-based planning
- introduction of principles of local economics, employment and ecology
- showing how the available land and water resources can best be utilized to achieve maximum and sustainable economic benefits
- farm diversification into commercial crops (e.g. fruits and vegetables)
- generation of increases in on-farm employment and incomes
- providing stimulus to downstream agro-industries
- the overall empowerment of rural women
- opportunities for skill development in seed production, horticulture, vegetables & poultry
- the promotion of micro-credit and savings programmes to generate capital to establish small rural enterprises
- introducing alternative fertiliser sources from productive sanitation systems to provide a resilient, readily available and cheap source of nutrients for this large stakeholder group of over 2 billion smallholder farmers
- development of water-harvesting practices
- development of sustainable agricultural practices

4.3. WHO guidelines for safe reuse

In 2006, WHO together with UNEP and FAO published updated health guidelines for the reuse of human excreta, wastewater and greywater which will help promote the further development of sustainable sanitation and ecological sanitation (ecosan). The main objective of these guidelines is to provide an integrated preventive management framework for maximizing the protection of human health and the beneficial use of important resources.

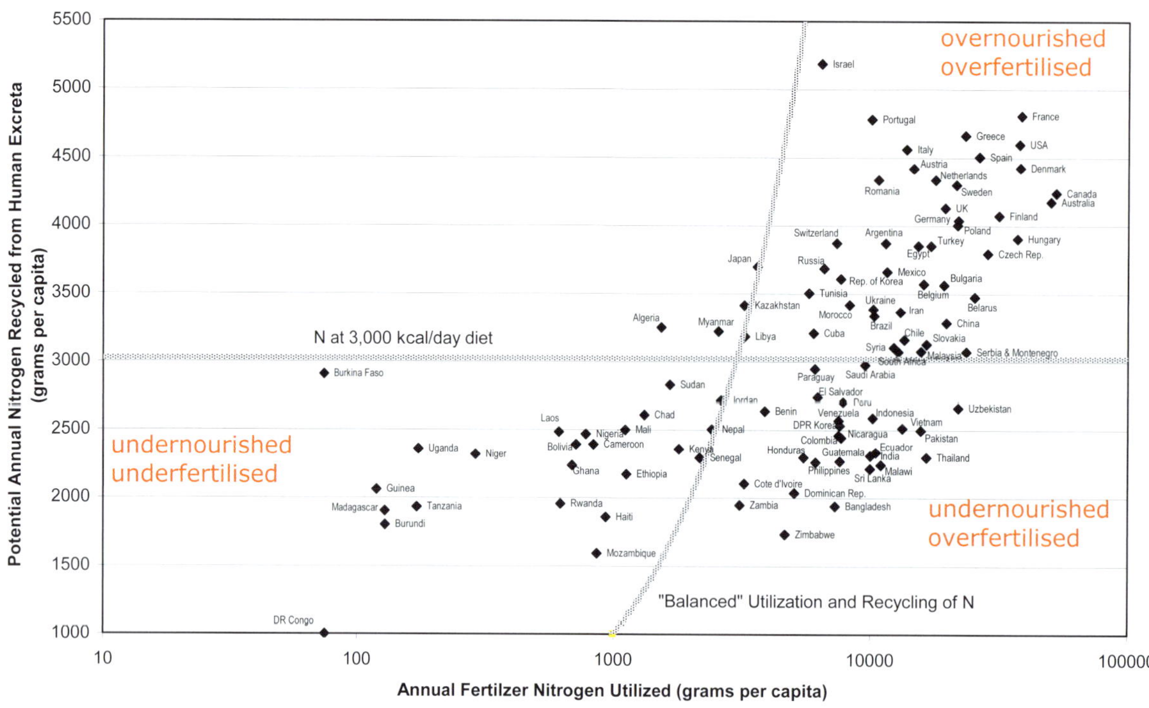

(Data source FAOstat 2005; SEI, 2005)

Figure 27: Comparison between chemical nitrogen fertiliser used and potential nitrogen fertiliser derived from sanitation systems

What is new in these guidelines is the break from the traditional practice of using isolated numeric monitoring parameters of pathogens as the sole guiding principle. The new guidelines use the so-called "Stockholm Framework" and promote the concept of health-based targets which can be achieved using a combination of risk-reducing options at various entry points from treatment of excreta, wastewater and sludge to measures taken in agricultural practice and finally at the level of the consumer. Although this all may seem very logical and obvious, there has not been until now the element of flexibility in the implementation of sanitation guidelines. It is now possible, for example, to use excreta or wastewater which are not yet fully treated if we can reduce the health risk later on in the agriculture, food production and consumption chain.

These guidelines are important in ensuring good sanitation practices and following microbial water quality standards. The guidelines are particularly directed to policy makers, people who develop standards and regulations, environmental and public health scientists, educators, researchers and sanitary engineers, and are intended to correct issues such as: irrigation or aquaculture using sewage; municipal or domestic wastes without substantial industrial inputs; faecal sludge derived from on-site sanitation facilities but not sludge produced from treatment of wastewater; and provide detailed information related to health protection – promoting good agriculture or aquaculture practices. There are four major explanations for the need to increase the use of excreta and greywater in agriculture and aquaculture:

- Increasing water scarcity and stress, and degradation of freshwater resources due to improper disposal of wastewater, excreta and greywater;
- Increasing human population and related increased demand for fertilizer to produce food and fibre;
- Increasing recognition of the resource value of wastewater and the nutrients it contains;
- The MDGs, with particular emphasis on the goals for ensuring environmental sustainability, health, and eliminating poverty and hunger.

Some of the most important specifications of the WHO guidelines are represented in Tables 3 to 9 below.

The new WHO guidelines for reuse are an important breakthrough in the process of introducing sustainable practices to the water and sanitation sectors around the world. Much remains to be done in order to implement these guidelines into national policies and regulations as well as standard extension practices. This will require enormous efforts in the area of outreach and capacity development and training targeted at the various governance and management levels. Nevertheless the ground has been laid for health-based targets and integrated risk management. These combined with policies and planning programmes bed well for positive developments in the coming decades ahead.

Table 3: Safety guidelines for large-scale treatment systems of greywater, excreta and faecal sludge for use in agriculture

(Source: WHO Guideline Executive Summary)

	Helminth eggs (no. per gm total solids or per Liter)	***E. coli*** (no. per 100 mL)
treated faeces and faecal sludge	<1/g total solids	<1000/g total solids
greywater for use in restricted irrigation	<1/Liter	$< 10^5$ (a) relaxed to $< 10^6$ when exposure is limited or regrowth is likely
greywater for use in unrestricted irrigation of crops eaten raw	<1/Liter	$< 10^3$ relaxed to $< 10^4$ for high-growing leaf crops or drip irrigation

(a)These values are acceptable due to the regrowth potential of E. coli and other faecal coliforms in greywater.

Table 4: Recommendation for storage treatment of dry excreta and faecal sludge before use at the household and municipal levels without adding new material

(Source: WHO Guideline Executive Summary)

Treatment	Duration	Comment
Storage; ambient temperature 2–20°C	1.5–2 years	will eliminate bacterial pathogens; regrowth of *E. coli* and *Salmonella* may need to be considered if rewetted; will reduce viruses and parasitic protozoa below risk levels. Some soil-borne ova may persist in low numbers.
Storage; ambient temperature >20–35 °C	>1 year	substantial to total inactivation of viruses, bacteria and protozoa; inactivation of schistosome eggs (<1 month); inactivation of nematode (roundworm) eggs, e.g. hookworm (*Ancylostoma/Necator*) and whipworm (*Trichuris*); survival of a certain percentage (10–30%) of *Ascaris* eggs (≥4 months), whereas a more or less complete inactivation of *Ascaris* eggs will occur within 1 year.
Alkaline treatment: pH >9	>6 months	if temperature >35 °C and moisture <25%, lower pH and/or wetter material will prolong the time for absolute elimination.

Table 5: Recommended storage times for urine mixture (1) based on estimated pathogen content (2) and recommended crops for larger systems (3)

(Source: WHO Guideline Executive Summary)

Storage temperature (°C)	Storage time (months)	Possible pathogens in the urine mixture after storage	Recommended crops
4	≥1	viruses, protozoa	food and fodder crops that are to be processed
4	≥6	viruses	food crops that are to be processed, fodder crops (4)
20	≥1	viruses	food crops that are to be processed, fodder crops (5)
20	≥6	probably none	all crops (6)

(1) Urine or urine and water. When diluted, it is assumed that the urine mixture has a pH of at least 8.8 and a nitrogen concentration of at least 1 g/l. (2) Gram-positive bacteria and spore-forming bacteria are not included in the underlying risk assessments, but are not normally recognized as a cause of any infections of concern. (3) A larger system in this case is a system where the urine mixture is used to fertilize crops that will be consumed by individuals other than members of the household from whom the urine was collected. (4) Not grasslands for production of fodder. (5) Not grasslands for production of fodder. (6) For food crops that are consumed raw, it is recommended that the urine be applied at least one month before harvesting and that it be incorporated into the ground if the edible parts grow above the soil surface.

Table 6: Health protection measures

(Source: Compiled from WHO Guideline Executive Summary)

Product consumers	Workers and their families	Local communities
excreta and greywater treatment	use of personal protective equipment	excreta and greywater treatment
crop restriction	access to safe drinking water and sanitation facilities at farms	limited contact during handling and controlled access to fields
waste application and withholding periods between fertilization and harvest to allow die/off of remaining pathogens	health and hygiene promotion	access to safe drinking water supply and sanitation facilities in local communities
hygienic food handling and food preparation practices	disease vector and intermediate host control	health and hygiene promotion
health and hygiene promotion	reduce vector contact	disease vector and intermediate host control
produce washing, disinfection and cooking		reduced vector contact

Table 7: Direct and indirect health effects of wastewater, excreta and greywater use

Direct health effects	Indirect health effects
direct outbreaks (developing and developed countries)	impact on the safety of drinking water, food and recreational water
contribution to background disease (e.g. helminths and other deadly diseases)	positive impacts on household food security and nutrition

Table 8: Monitoring and system assessment

	Validation	Operational monitoring	Verification
Function	proves that the system is capable of meeting its design requirements	provides information regarding the functioning of individual components of the health protection measures	takes place at the end of the process to ensure that the system is achieving the specified target
Purposes	performed when a new system is developed or when new processes are added	used on a routine basis to indicate that processes are working as expected	used to show that the end product (e.g. treated excreta or greywater; crops) meets treatment target and ultimately the health-based targets

Table 9: Important aspects and considerations of the WHO Guidelines

(Source: Compiled from WHO Guideline Executive Summary)

Important aspects	Comments
Socio-cultural	Human behavioural patterns are a key determining factor in the transmission of excreta-related disease. The social feasibility of changing certain behavioural patterns in order to introduce excreta or greywater reuse schemes or to reduce disease transmission in existing schemes needs to be assessed on an individual project basis. Cultural beliefs and public perceptions of excreta and greywater reuse vary so widely in different parts of the world that one cannot assume that any of the local practices that have evolved in relation to such use can be readily transferred elsewhere.
Economic and financial	Economically worthwhile projects can fail if the financial plan is not well made. Economic analysis and financial considerations are crucial for encouraging the safe use of excreta. Economic analysis seeks to establish the feasibility of a project and enables comparisons between different options. The cost transfer to other sectors such as the health and environmental impacts on downstream communities also needs to be included in cost analysis. The ability to profitably sell products fertilized with excreta or irrigated with greywater also needs analysis.

Important aspects	Comments
Environmental	Excreta are an important source of nutrients for many farmers. Urine alone contains more than 50% of phosphorus excreted by humans. The direct reuse of excreta and greywater on arable land tends to minimize the environmental impact in both the local and global context. Reuse of excreta on arable land secures valuable fertilizers for crop production and limits the negative impact on water bodies. The nutrients may, however, percolate into the groundwater if applied in excess or flushed into the surface water after excessive rainfall. Phosphorus is an essential element for plant growth. World supplies of accessible mined phosphate are diminishing. About 25% of mined phosphorus ends up in aquatic environments or buried in landfills or other sinks. This discharge into aquatic environments causes eutrophication of water bodies. The diversion and use of urine in agriculture can aid crop production and reduce the costs of and need for advanced wastewater treatment processes to remove phosphorus from the treated effluents.
Policy	In developing a national policy framework to facilitate the safe use of excreta as fertilizer, it is important to define the objectives of the policy, assess the current policy environment and develop a national approach. National approaches need to be adapted to the local socio-cultural, environmental and economic circumstances, but they should be aimed at progressive improvement of public health. Interventions that address the greatest local health threats first should be given the highest priority. Appropriate policies, legislation, institutional frameworks and regulations at the international, national and local levels facilitate safe excreta and grey-water management practices. Yet these frameworks are lacking in many countries where excreta and greywater management is practiced.
Planning and implementation	Planning and implementation of programmes for the agricultural use of excreta and greywater require a comprehensive, progressive and incremental approach that responds to the greatest health priorities first. This integrated approach should be based on an assessment of the current sanitary situation and should take into account the local aspects related to water supply and solid waste management. A sound basis for such an approach can be found in the Bellagio Principles which prescribe that stakeholders be provided with the relevant information, enabling them make informed choices. In addition, project planning requires consideration of several different issues, identified through the involvement of stakeholder applying participatory methods and considering treatment, crop restriction, waste application, human exposure control, costs, technical aspects, support services and training both for risk reduction and for maximizing the benefits from an individual as well as a community point of view.

5. Policy, Planning and Implementation

Although it seems natural that sanitation implementation in both rural and urban situations should be initiated based on sound policies, safety guidelines and comprehensive planning involving the key stakeholders, this is not the general or common case. Often inappropriate choices are made because the policies don't exist or are not applied and planning is inadequate. Since sanitation is not normally a "political issue" for a community and since private and public dialogues are lacking in almost all cases around the world, the sector is tainted by dysfunction, with inadequate political leadership and institutional follow-up. Often it is left to public health inspectors and engineers to provide services that stakeholders have little knowledge or understanding about. And more often than not the services provided are inadequate especially in developing countries.

5.1. Sanitation policy

The role of safe sanitation in alleviating poverty and improving health is widely recognized and reflected in the MDG target of halving the proportion of people without sustainable access to basic sanitation by 2015. One possible reason why sanitation provision continues to lag far behind that of other services is the absence of effective sanitation policies which makes governments invest far less in sanitation provision than in water services. Shrestha *et al.* (2005) state that policy and guideline development and formulation should: be realistic and relevant; be informed-based; develop local targets; create informed demand; enable a role for the private sector; and improve coordination. A poor policy can constrain action while a good policy can facilitate change and development. Good policy provides the framework within which those who are seeking to improve sanitation can operate. It enhances understanding of sanitation-related issues, set clear overall objectives, clarify responsibilities and provide incentives for action to achieve the objectives. All these help to establish an environment in which sanitation can be taken seriously and, therefore, addressed on a scale that can significantly contribute to improved national health, well-being and economic development opportunities.

Most countries have combined national water supply and sanitation policies. According to Tayler and Scott (2005), a combined policy can take account of the strong links between water, sanitation and health. Outlaw *et al.* (2007), however, state that linking water and sanitation as one sector undermines support for sanitation as represented by funding and policy support decisions. Thus, Tayler and Scott (2005) insist that if sanitation is to be given due attention, it needs its own policy and this is more likely to happen if sanitation is covered by key general policy instruments such as the Poverty Reduction Strategy Papers (PRSPs). Most combined policies focus on water supply and deal with sanitation in a rather perfunctory way.

Box 1: Key elements of national sanitation policies

(Adapted from Elledge et al.,2002; Tayler and Scott, 2005)

- political will (the support given by politicians, officials and other influential people or organisations)
- acceptance of policies (including its relevance to stakeholders)
- legal framework (existence and relevance of laws, acts and regulations)
- population targeting (consideration given to the needs of the urban poor, residents of small towns, refugees, displaced persons, women, etc.)
- adequate level of service
- consideration given to health, the environment and financial issues
- institutional roles and responsibilities

For instance, Nepal has had a separate national sanitation policy since 1994. More recently, the Rural Water Supply and Sanitation National Policy, Strategies and Sectoral Strategic Action Plan - 2004 (RWSSNPSSSAP-2004), the integrated policy for both sanitation and water supply, was developed by His Majesty's Government of Nepal (HMGN). This was followed by development of National Guidelines for Hygiene and Sanitation Promotion - 2005 within the integrated policy framework. Overall policy and guideline objectives are framed in terms of the sanitation coverage to be achieved and the institutional arrangements for implementing policy. There is scope for improvements in the policy content, but a key challenge in Nepal is how to effectively implement policy. This is undoubtedly one of the reasons why infant and under-five mortality rates remain high throughout Nepal with an estimated 15,000 children dying each year due to diarrhoeal diseases caused by poor environmental sanitation and lack of access to quality water supply (Shrestha *et al.* 2005). In addition, Tayler and Scott (2005) report that Nepal's Rural Water Supply and Sanitation Policy (RWSSP) of 2004 focuses strongly on community management, but has little to say about the role of individual households which is likely to be significant, particularly in rural sanitation provision and management. The assessments of Tayler and Scott (2005) show that in both Ghana and Nepal, existing policy targets for sanitation coverage are higher than those included in the MDGs. In Nepal, the target set in the RWSSP of 2004 is to achieve 100% sanitation coverage by 2017. Whereas in Ghana, the National Environmental Sanitation Policy (NESP) includes the target that at least 90% of the population should have access to an acceptable domestic toilet while the remaining 10% should have access to hygienic public toilets. Far greater financial resources than currently allocated are required if these ambitious targets are to be met in both countries (See Box 2).

Box 2: Shortfalls in funding in Nepal and Ghana

(Adapted from Tayler and Scott, 2005)

Nepal: To reach the sanitation target, an additional 14,000 households need to be served per month between 2000 and 2015, and an additional 11,300 households need to be served per month to reach the drinking water target. The total financial requirement to meet these targets is US$ 1,087 million and the resource availability for 2000 to 2015 is US$ 755 million, resulting in an annual resource gap of US$ 23 million. The shortfall in funds is estimated to be over 50% (US$6 million per year) of the level required to meet the MDG sanitation target, let alone the more ambitious target set out in national policy (WaterAid, 2004).

Ghana: Given the current levels of investment in Ghana, it is estimated that rural sanitation coverage would actually fall from around 31% to about 24%. This is based on figures in a report prepared by Lukman Salifu in Ghana as part of joint WEDC/WaterAid research on sanitation policy in Ghana.

5.2. Sanitation planning systems

The sustainability of water and sanitation projects has, historically, not been satisfactory. Of the water supply and sanitation projects evaluated by the World Bank in 2001, only 50-66% was deemed to be satisfactory and less than half were rated likely to be sustainable (World Bank, 2003). Project assessments consistently report cultural constraints, behavioural change, prohibitive costs, lack of political and managerial support, or low community demand as reasons for low success rate of water and sanitation projects. This wide array of constraints needs to be overcome if sanitation is to be brought to scale in a sustainable manner.

The sanitation intervention starts with the planning phase. It is thus important to keep all constraints against successful sanitation interventions in mind already at the sanitation planning stage. Participatory and holistic approaches to sanitation planning can increase the potential for a sustainable system through better management of the numerous risk-factors and capacity development within the local domains for successful operation and maintenance of the systems. In recognition of this, a number of organizations have developed and are promoting planning frameworks for sanitation based on the assessment of user priorities at different levels of decision-making within the urban environment (EAWAG, 2005; IWA, 2006; Ridderstolpe, 2000, NETSSAF, 2008; Kar, 2005; Kar and Chambers, 2008; Wood *et al.* 1998). Users of these methods can then select appropriate systems that will satisfy the functional requirement of the various stakeholders.

There have been numerous support tools and frameworks designed to aid in planning processes for sanitation/wastewater planning and management. While methodologies vary in their emphasis, both whether they are geared towards more rural or urban areas, and whether or not they are using top-down or bottom-up planning techniques, there is a growing consensus on the need to include stakeholders in the planning and implementation processes.

Most sanitation planning tools are either geared towards urban environments (e.g. HCES, Sanitation 21) or more rural settings (e.g. CLTS, PHAST). In an urban/peri-urban/small town setting, planning tools are often adapted from rural settings (e.g. open defecation behaviour) and the non-served percentage of the population has little faith in service delivery. The feasible sanitation systems might thus be decentralized in character making it difficult to implement top-down urban planning tools and to apply participatory planning methods. Urban-oriented planning tools are often weak in steps creating sanitation demand and stimulating the necessary behaviour change, which is paramount when planning for sanitation in areas where open defecation is highly prevalent. On the other hand, participatory, rural development tools (such as PHAST) are designed to work in areas where centralized regulations on health and the environment might not be applied, which makes them more difficult to use without adaptation in peri-urban areas. The peri-urban area will certainly be subject to both physical planning, sanitation and environmental regulation and by-laws, and space constraints among other things, which will lead the sanitation development in a way that might not be completely compatible with a method such as the CLTS. However, the elements of CLTS (e.g. the participation and ignition) will still be necessary to motivate behavioural change to stop open defecation.

Table 10: An overview of different sanitation planning tools

(adapted from McConville, 2008 and NETSSAF, 2008)

	PHAST (Wood et al. 1998)	Open Planning of Sanitation (Ridderstolpe, 2000)	Strategic Choice Approach (Friend, 1992; Wright, 1997)	HCES (EAWAG, 2005)	Sanitation 21 (IWA, 2006)	MCDSS (Wiwe, 2005)	Guidelines for Municipal Wastewater Management (UNEP et al 2004)	CLTS (Kar, 2005)	NETSSAF (2008)
Problem Identification	problem identification	problem identification	**shaping** of problem structure	request for assistance launch of planning process	Stakeholder Identification assessment of external factors driving decision	definition of problem, goals and objectives	monitoring assessment and identification of need for action review of information	introduction and rapport building participatory analysis ignition moment	project start-up and launch of the planning process creation of a demand for improved sanitation
Define Objectives	problem analysis	identification of boundary conditions terms of requirement		assessment of current status assessment of user priorities	assessment of management capacity within city for any system identification of interests and objectives of stakeholder groups	definition of criteria	identification of needs and opportunities setting objectives and standards		assessment of existing sanitary situation and user priorities
Design Options	planning for Solutions	analysis of possible solutions	**designing** available courses of action	identification of options	technical analysis of existing and potential systems identification of alternative systems, including management requirements	definition of alternatives meeting criteria			construction of demonstration units
Selection Process	selecting options	choice of the most appropriate solution	**comparing** actions against significant criteria + **choosing** the preferred alternative and how to reach it	evaluation of feasible service combinations	assessment whether proposed systems meet objectives and management requirements	definition of preferences decision making based on alternatives and preferences			identification of feasible sanitation concepts and service systems
Action Plan for Implementation	planning for new facilities and behaviour change			consolidate plans finalize plans implementation			formulation of management plan design management tools operational management institutional arrangements formal adoption	action planning by community	consolidation and finalization of sustainable sanitation plans implementation
Monitoring & Evaluation	planning for monitoring and evaluation participatory evaluation			monitoring, evaluation and feedback			operational management of water quality evaluation	follow-up	participatory monitoring and evaluation

Box 3: Generic planning steps

(Excerpt with adaptations from McConville, 2008)

Step 1: Problem Identification

This step defines the context of the current situation and the scope of the problem to be addressed. It is the core of the first question in strategic planning, "Where are we now?" It requires an understanding of the existing sanitation structures, as well as, stakeholder attitudes and institutional realities. Here planners also need to identify external and internal risk factors and assumptions. Useful tools during this stage include Political, Economic, Social, and Technological (PEST) and Strengths, Weaknesses, Opportunities, and Threats (SWOT) analyses (Örtengren, 2004).

Step 2: Define Objectives

This step defines a vision of the future by answering the question "Where do we want to go?" It requires participatory approaches to really identify the interests and priorities of the various stakeholders. Planners will need to recognize potential conflicting interests and competing priorities between interest groups, a process which will make sure that the competing priorities are made visible in the planning process and can be addressed. Reaching an acceptable consensus on objectives often requires compromise and equitable treatment of all interest groups (Hajer & Wagenaar, 2003). Issues important for sustainability, in the given planning setting, will have to be considered in the formulation of the objectives of the system, if sustainable sanitation is to be achieved. The objectives should also be transparent, measurable and clearly stated. Useful tools during this step include participatory assessments as tools, such as those used in Participatory Hygiene And Sanitation Transformation (PHAST) and in Community-Led Total Sanitation (CLTS), and in setting the Terms of Requirements (ToR) for a sanitation system, (Kvarnström & af Petersens, 2004).

Step 3: Identify Options

The next three steps work to answer the question of "How do we get there?" The first part of this is to identify possible solutions. This step relies heavily on the principle of technical flexibility in order to generate a wide range of potential solutions. Potential options should be generated based on a systems perspective of required functionality. Therefore, both centralized and decentralized systems with the potential to meet the objectives should be considered (Ridderstolpe, 2000). It may also be possible to mix technologies that serve different demographic domains, or different waste flows (i.e. greywater, urine diversion, and solid waste), (IWA, 2006). There are a variety of sanitation technology databases that can be explored during this phase for idea generation (i.e. SANEX™ Version 2.0, NETSSAF, 2008, Tilley *et al.*, 2008)

Step 4: Selection Process

The selection process includes feasibility studies and critical comparison of the potential solutions. The chosen solution must be matched to technical objectives, affordability, and managerial capacities in the local context. A variety of analytical and decision-support tools exist to aid in these feasibility assessments (e.g. LCA and EIA). Multi-Criteria Decision Support Systems (MCDSS) are also commonly used when there is a need to identify trade-offs between of a variety of information, often including both quantitative and qualitative data, as is the case with sanitation. The advantages of using MCDSS in decision-making are that it can increase transparency, stakeholder participation, and optimization by application of several criteria in the decision process (Wiwe, 2005). The selection process should also be a participatory process and include stakeholder input on potential designs.

Step 5: Action Plan for Implementation

This step is not explicitly stated in all planning frameworks, however it is the core outcome of the previous steps as it translates the decision process into a direct plan on how to reach the agreed objectives (Örtengren, 2004). The action plan is the actual planning document which details how to implement the chosen technologies/systems and supporting capacity building exercises. It will include a timeframe for objectives to be met, as well as the roles and responsibilities of the stakeholders.

Step 6: Monitoring and Evaluation

Ensuring the success of a sanitation system requires feedback loops and monitoring the system so as to be able to correct any problems that arise (Mugabi *et al.*, 2007). Mechanisms for monitoring and evaluation should be discussed and implemented during the planning process. This includes identifying measurable indicators for monitoring, but also a system of incentives to encourage investment in sanitation and ownership of the outcomes.

Table 10 gives a brief overview of the planning steps of some of the different planning tools. The planning steps have been sub-grouped under six different generic planning steps: (i) problem identification; (ii) defined objectives; (iii) design options; (iv) selection process; (v) action plan for implementation; and (vi) monitoring and evaluation.

The planning frameworks cover most of these different generic planning steps but differ in the level of relative importance. It can be noted that some planning methods stop when the choice of system is made and do not have steps geared toward action planning for implementation and evaluation and follow-up. An overview of the different generic planning steps has been presented by McConville (2008), see Box 3.

5.3. Implementing improved sanitation

A critical component of policy and planning is the scope provided by a country in its attempt to increase coverage of improved sanitation. The JMP (WHO/UNICEF, 2008a) provides a definition for improved sanitation to include facilities that ensure hygienic separation of human excreta from human contact. These include:

- flush or pour-flush toilet/latrine to:
  - piped sewer system
  - septic tank
  - pit latrine
- ventilated improved pit (VIP) latrine
- pit latrine with slab
- composting toilet

Exactly what is meant by composting toilet is not spelled out but this should include such things as Arborloo, Fossa alterna, single and double vault urine-diverting toilets, Clivus multrum and similar toilets. The element of containment should be a critical component of the working definition. The definition should also be restrictive for cases where a flush toilet empties untreated directly into water that is used for drinking purposes but this is not yet the case. The result is that the coverage data published by the JMP based on household survey data and census data may not be an accurate reflection of the real situation regarding sanitation functionality and health and environmental safety. Discussions are ongoing between the WHO, UNICEF and other development partners to introduce the "next generation" of monitoring systems which will incorporate the aspects of improved management and sustainability and not just the enumeration of existing installations (eg World Water Week, 2008).

The case of Uganda

An example of how difficult it is to put these definitions into practice is the case for Uganda. Uganda has well-written national policies, but the challenge becomes more obvious when it comes to implementation in a decentralized governance environment. In general, the major difficulty is in creating an environment in which national policy is implemented at the lowest level of government. It is quite common to find local government offices lacking the technical, managerial, and financial capacity to address the real sanitation needs. Programmes also tend to focus on facilities and give less attention to software such as health and hygiene promotion. This is where ecological sanitation that demands a commitment to practise may find problems in implementation (Johansson and Kvarnström, 2005). Uganda had the highest rate of sanitation coverage in Africa in the 1960s, with over 90% of households having their own latrine.

However, in the ensuing decades of dictatorship and civil unrest these figures dropped dramatically, resulting in significantly lower coverage rates today. Current coverage figures vary by data source. The demographic health survey (DHS) states that 'improved' household sanitation coverage in rural areas is just 8%; the Uganda district health inspector (2007) cites national coverage at 57%. The discrepancies in these coverage figures appear to relate primarily to differences in the definitions of 'safe/improved sanitation' (Table 11). For example, within Uganda, the Environmental Health (EH) Policy includes pit latrines that meet performance-based standards within coverage figures. However, the Uganda Bureau of Statistics (UBOS) in analyzing 2006 DHS data, classifies pit latrines as unimproved if they do not have a concrete slab, thereby omitting them from coverage rates.

Table 11: Differences in the definition of improved sanitation in Uganda

(Source: Outlaw et al., 2007)

Ugandan Bureau of Statistics (UBO)/ Demographic Health survey (DHS)	Environmental Health Division (MOH)
Technology-based	**Performance-based**
• flush toilet • Ventilated improved pit latrine (VIP) • pit with concrete slab • composting latrine	• must provide privacy • faeces cannot be less than 3 feet from top of latrine pit • slab must be structurally safe but can be made of wood

Given that the DHS records report that 41% of rural households are using unimproved pit latrines without concrete slabs, this could explain most of the difference between DHS and EH indigenous latrine coverage figures. The Ugandan government's overall policy objectives for the water and sanitation sector are 'to provide sustainable provision of safe water within easy reach and hygienic sanitation facilities…to 77% of the population in rural areas and 100 percent of the urban population by the year 2015 with an 80% - 90% effective use and functionality of facilities' (MOW/Environment, 2007).

The case of Ethiopia

Like many other countries, the Ethiopian government has always prioritized, and is still prioritizing, water supply over rural sanitation and hygiene promotion, and as such much more investment has been made in the water supply sector than in the rural sanitation and hygiene promotion sector. The Ethiopian definition of improved sanitation draws on the definitions of WHO and UNICEF while emphasising the key principle of 100% improvement. 100% adoption of improved sanitation and hygiene is the process where people demand, develop and sustain a hygienic and healthy environment for themselves by erecting barriers to prevent the transmission of diseases, primarily from faecal contamination (Box 4).

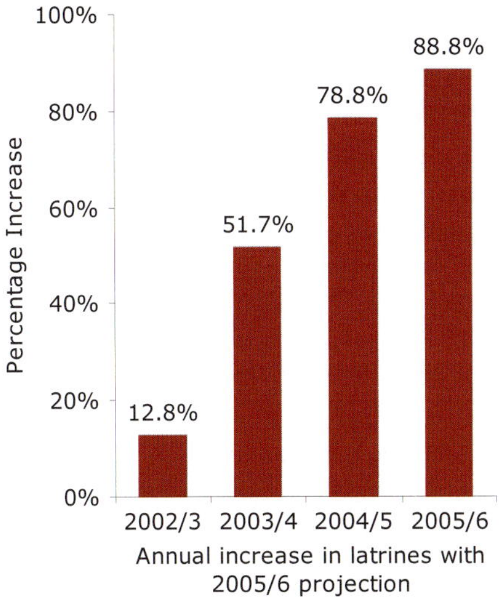

(Adapted from WSP Field Note January 2007. "From Burden to Communal Responsibility: A Sanitation Success Story from Southern Region in Ethiopia")

Figure 28: Rapid increase in sanitation coverage in the southern region of Ethiopia stimulated by activities lead by the Regional Health Bureau

The Christian Relief Development Association reported in 1999 that 75% of patients seeking care at hospitals in urban areas in Ethiopia were suffering from water- and excreta-related diseases (CRDA, 1999). Although there are regional variations in latrine status in Ethiopia, Teka and Mulugeta (2003) report that some kind of latrine access ranges between 9% in rural areas to 72% in the urban areas, giving a national average of 18% which is mainly traditional latrines made from locally available materials. In 1994, census showed that only 6% of rural households had latrines. By the year 2000, 18% of rural households nationwide had a sanitation facility in or near their homes, and 72% of urban households had a facility (WS Atkins et al., 2003).

Recent evidence has shown a new trend break in that the southern region of Ethiopia has experienced unprecedented significant growth in sanitation coverage in the past few years (Figure 28) thanks to strong leadership from the Regional Health Bureau. Significant change can occur.

5.4. Choosing appropriate systems

One of the major problems in meeting the MDG sanitation target and the WHO/UNICEF target on "Sanitation for All" by 2025 is that far too few professionals (in both developing and industrialized countries) are aware of all the available sanitation options and are able to select the most appropriate sanitation arrangement for any given community (Mara *et al.*, 2007). This combined by the fact that there is limited dialogue and knowledge on the part of the stakeholders leads to inadequate planning procedures. Choosing a sanitation system that is appropriate for a community must take into account several basic factors including: behavioural and cultural acceptability; ease of use; user safety; and cost.

These "standard" criteria are mainly built around the needs of the toilet user and his/her ability to pay. But they are also based on the common assumptions surrounding human behaviour of "flush and forget" for waterborne systems or "hide and forget" for dry systems using deep pit latrines. The *sanitation system* is not under public scrutiny but is left to a specialist group of engineers challenged with the task of making the excreta and used water "disappear" at the lowest possible cost. As a result, deep pit latrines (e.g. 5 meters deep) are not designed for emptying, buried septic and conservancy tanks and sewerage pipes cannot easily be monitored to check if they are leaking, seepage pits are covered with a concrete seal never to be monitored or maintained, and the list goes on. Maintenance, in general, is not a top design priority for both communal and household installations. Most conventional sanitation systems sooner or later become dysfunctional, calling for mammoth investments in upgrading of hardware. This is a world-wide problem and not at all the subject of public scrutiny. When it comes to advanced sewage treatment plants, the capacity to maintain and manage these facilities is particularly lacking in the developing countries.

In order to break this trend it is necessary go beyond the conventional planning approaches and include other aspects related to sustainability. These include: the ease and affordability of systems maintenance; the level of containment and ensuing environmental safety; the system's energy and resource efficiency; and the potential for reuse in agriculture.

This process of broadening the "playing field" for sanitation has only commenced in the past few years and the level of professional capacity to carry this out is extremely limited. In particular there is an acute need for more comprehensive planning instruments that also involve stakeholders and provide information on the various options available for a particular community. Professional capacity development is the other major area requiring attention whereby knowledge about sustainable approaches can be shared.

Table 12 illustrates how diverse the options for technology can be. They need to be divided into the main subsystems: toilet or collection, on-site storage and treatment, transportation, off-site storage and treatment, reuse and disposal. The pioneer work which was part of the EU-sponsored NETSSAF project produced detailed overviews of the various sanitation options and did comparisons as to level of appropriateness for West Africa (NETSSAF, 2008).

In addition an attempt has been made in Table 13 to provide a template in order to compare the various sustainability criteria for a selection of common sanitation systems both offsite and onsite, waterborne and dry. Departing from a conventional base case, this allows one to examine alternatives in the process of deciding on what sanitation systems are most appropriate using various criteria such as: ease of use; social-cultural acceptability; system affordability; ease of system maintenance; risk to the user and the environment (i.e. downstream); reuse potential; and resource and energy efficiency. Criteria for such a planning exercise are to be generated locally for the particular context and these provided here should be seen as possible suggestions.

Box 4: The sanitation vision for Ethiopia

Source: Ministry of Health, Federal Democratic Republic of Ethiopia, 2005

100% adoption of improved (household and institutional) sanitation and hygiene by each community which will contribute to better health, a safer, cleaner environment and the socio-economic development of the country.

Conditions for Success

- getting consensus that the current limited and inappropriate access to sanitation and hygiene is a problem.
- ensuring dedicated political commitment, support and action.
- achieving accountability through 'minimum' performance contractual agreements at all levels.
- gaining inter-sectoral collaboration using convincing promotion of the benefits while emphasising the risks.
- allowing for minimum contact time of health extension workers (guidance and health education) with households.
- realizing community empowerment and responsibility through using viable local solutions.
- implementing effective supportive supervision and monitoring processes which are linked to performance contractual agreements.

The Three Strategic Pillars for Improved Sanitation and Hygiene

Pillar 1: an enabling framework to support and facilitate an accelerated scaling-up through policy consensus, legislation, political commitment, inter-sectoral co-operation, partnership, capacity building linked to performance contractual agreements, supportive supervision, research and monitoring

Pillar 2: sanitation and hygiene promotion through participatory learning, advocacy, communication, social marketing, incentives or sanctions to create demand and forge behavior change

Pillar 3: improved access to strengthen the supply of sanitation through appropriate technology solutions, product and project development, and support to local producers and artisans

Why is a Strategy Necessary?

- in Ethiopia more than 250,000 children die every year from sanitation- and hygiene-related diseases
- some 60 percent of the disease burden is related to poor sanitation and hygiene
- a low number of households (between 6 and 18 percent) have access to improved sanitation
- less than 1 percent of the health budget is dedicated to sanitation and hygiene improvement
- the annual sanitation 'fall-out' costs are devastating

What are the Benefits of Improved Sanitation?

- health: diarrhoea prevention, mortality decreased, curative care reduced and nutrition improved
- socio-economic: fitter workforce, less time caring for the sick, less money spent treating sicknesses
- educational: enhanced school attendance by girls and attained higher levels of education
- social – privacy, dignity, safety and a cleaner environment
- gender – women stand most to gain from improved sanitation and hygiene benefits
- political – women represent 50 percent of the electorate, making sanitation an important political issue

How Big is the Challenge?

- the 2015 MDG is to halve the proportion of Ethiopians without access to improved sanitation which is equivalent to 11 million households in 11 years (a million households per year).
- 100% sanitation and hygiene means SOME FOR ALL not more for some.
- the imbalance between curative and preventive health care requires raising the sanitation profile.

Where Are the Opportunities?

Positive examples from the regions and zones offer lessons on the scale that can be achieved such as:

- the Southern Nations and Nationalities People's Regional State (SNNPRS) has achieved 65% latrine coverage with their own resources through political (and budget) commitment, inter-sectoral collaboration, accountability and community ownership
- there is increasing inter-sectoral convergence around a single sanitation strategy
- the new Health Service Extension Programme (HSEP) has seven dedicated sanitation packages
- the WASH Movement is gathering momentum through its focus on hand washing in 2004
- an advancing decentralisation is dedicated towards people taking ownership of their own development
- donors are recommending funds to be dedicated for sanitation and hygiene in WASH Programmes
- National Hygiene and Sanitation Strategy to enable 100% adoption of improved hygiene and sanitation

Table 12: Sanitation technologies

(modified from NETSSAF, 2008)

Toilet & collection technologies
- Cistern-flush toilet
- Low-flush toilet
- Pour-flush toilet
- Urine-diversion toilet
  -flush toilet
  -waterless toilet
- Urinal
  -waterless urinals
  -low-flush urinals
- Dry toilet squatting slab
- Simple pit latrine
- Ventilated improved pit latrine (VIP)
- Double pit latrine
- Double vault latrine
- Composting toilet
  -shallow pit
  -vault
  -Arborloo latrine
  -Fossa alterna

Transport technologies
- Gravity sewers
- Small bore sewers
- Simplified sewerage
- Vacuum sewerage
- Open drains
- Urine pipes
- Manual urine transport
- Trucked urine transport
- Manual or suction truck faecal sludge emptying and transport

On-site storage and treatment technologies

Related to wastewater
- Septic tank
- Cesspit
- Anaerobic baffled reactor
- Anaerobic digester
- Trickling filter
- UASB reactor (Upflow Anaerobic Sludge Blanket)

Related to urine
- Urine long-term storage
  -in different types of containers
  -in large storage tank
- Urine can, bucket or container storage
- Urine desiccation

Related to excreta and faecal sludge
- Faecal sludge co-composting
- Faecal sludge treatment by
  -constructed wetlands (humification)
  -unplanted drying beds
  -settling ponds
  -anaerobic digestion

Related to greywater
- Greywater pre-treatment (screens, seals, filters)
- Flotation – grease trap
- Slow sand filtration
- Horizontal subsurface flow constructed wetland
- Horizontal free flow constructed wetland system
- Vertical flow constructed wetland system
- Greywater garden (mulch trench)
- Green walls/Tower garden
- Subsurface wastewater infiltration system
- Anaerobic filtration

Off-site treatment technologies

Related to wastewater
- Pre-treatment
- Waste stabilization ponds
- Advanced Integrated Pond Systems
- Floating macrophyte ponds
- Constructed wetlands
- UASB technologies
- Conventional activated sludge systems
- Integrated Fixed-film Activated Sludge
- Membrane biological reactors

Related to urine
- Off-site urine storage tank
- Urine MAP-dissipation

Reuse technologies
- Urine direct application
- Urine on-site reuse
- Urine mechanized off-site reuse
- Faecal sludge & excreta use in agriculture
- Effluent (wastewater) application in agriculture
- Effluent (wastewater) and faecal sludge (excreta) use in aquaculture

Disposal technologies
- Soakaway pit
- Infiltration trench/field

Table 13: Examples of two typical conventional base cases for on- and offsite sanitation systems with examples of both wet and dry improved sanitation options. The various sustainability criteria are listed as a template in order to arrive at possible improvements to the base case.

	On-site system with separate greywater treatment and discharge (e.g. infiltration bed)							
	dry systems ←			**base case**	**→ pour-flush/flush systems**			
	ventilated dry-vault toilet with urine-diverting squatting pan or seat riser; faeces composted; agro-reuse of products	soil composting shallow pit latrine (Arborloo or Fossa alterna) with slab; option of squatting pan or seat-riser; option of urine diversion; periodic emptying and agro-reuse	ventilated improved pit latrine (VIP) with slab; periodic emptying and landfill disposal	deep pit latrine with slab; periodic emptying and landfill disposal	wet pit latrine with toilet fixed to slab; periodic emptying landfill disposal	toilet with direct discharge to cess pit	toilet with direct discharge to septic tank and effluent drains to seepage bed	toilet with direct discharge to single unit biogas fermentor ; periodic deludging, composting and agro-reuse
	Off-site system with the choice of mixed or separate black and greywater, dry collection or compost							
	dry systems ←			**base case**	**→ pour-flush/flush systems**			
	ventilated urine-diverting dry toilets with collection of faeces in bins for community composting; urine stored in tanks; agro-reuse of products; greywater discharged by small-bore pipe to constructed wetland	ventilated dry squatter or seat-riser toilets with composting bin (eg Clivus multrum); periodic emptying of compost; agro-reuse; greywater discharged by small-bore pipe to constructed wetland	ventilated dry squatter or seat-riser toilet connected to bucket or conservancy tank; periodic collection and transport to stabilisation pond or landfill; greywater discharged to open drain	pour-flush/flush toilets and greywater piped directly to open street drain leading to surface discharge at city limits	toilet connected to conservancy tank; blackwater collected and transported to stabilisation pond - desludged and composted locally; agro-reuse; greywater discharged to open drain	toilet blackwater and household greywater connected to septic tank; effluent drained to small-bore pipe sewerage and discharged to constructed wetland	toilet blackwater and household greywater connected to large-bore sewer with pond or wetland treatment and surface discharge	toilet blackwater and household greywater connected to large-bore sewer with advanced sewage treatment plant and surface water discharge
criteria								
risk to user								
ease of use								
environmental safety								
risk during maintenance								
affordability								
treatment								
reuse								
resource efficiency								

6. Costing the Scaling-up to Meet the MDG Target

6.1. Cost-benefit analyses

Estimates of the costs to achieve worldwide coverage of water supply and sanitation by 2025 are between US$11 – US$132 billion per year. Bos *et al.* (2005) estimate a required per capita investment of US$34 per year, amounting to a required worldwide investment of US$200 billion per year (assuming that conventional technologies are used). These cost estimates do not include full coverage of wastewater treatment (estimated at US$70 – US$90/cap/y), which would raise the total worldwide annual investment requirement to some US$600 – US$800 billion/year. Some studies have estimated the cost of attaining the MDG target for sanitation and indicate amounts in the range of US$9 billion to US$30 billion per year at the global level.

An explanation for the wide range of cost to attain the MDG sanitation target is the different methodologies used, the various unit cost assumptions, technology choice and definitions of adequate service levels. The per capita cost of improved sanitation is higher compared to the per capita cost of improved water supply services. This is because the basic water supply services are mainly public and shared by a higher number of persons as opposed to basic sanitation options. Similarly, the absolute number of persons needing access to improved sanitation to meet the MDG target is far higher than the number of persons needing access to improved water supply. Halving the proportion of people without sustainable access to improved water supply would cost around US$1.7 billion per year. Halving the proportion of people without sustainable access to both improved water supply and improved sanitation would cost around US$11.3 billion annually (Hutton and Haller, 2004).

Hutton *et al.* (2004) also show that achieving the global MDG target in water and sanitation would bring substantial economic gains from both health and other benefits i.e. each US$1 invested would yield an economic return of between US$3 and US$34, depending on the region. The benefits would include an average global reduction of diarrhoeal episodes of around 10%. If the goal for water and sanitation were met, the health-related costs avoided would reach US$7.3 billion per year, and the annual global value of adult working days gained as a result of less illness would be almost US$750 million (Bartram *et al.*, 2005).

According to estimates of Evans *et al.* (2004), (Table 14) the total cost to meet the MDG target would range from an annual cost of US$3.1 billion by using the simplest possible approaches to US$80 billion by using the most expensive technologies including tertiary wastewater treatment. Approximately US$ 10 billion per year would be required to supply low-cost water and sanitation services to people who are not currently supplied, and a further US$15 to US$20 billion a year to provide them with a higher level of service and to maintain current level of service to people who are already supplied (Toubkiss, 2006).

Box 5: Health benefits of improved sanitation and hygiene

(Source: Esrey et al (1991); Ethiopian National Hygiene and Sanitation Strategy, 2005)

- Pit latrines, when used by adults themselves and for the disposal of infant's stools, can reduce diarrhoea by 36 percent or more, cholera by 66 percent, and worm infestations by between 12 and 86 percent.
- Hand washing with soap (or a substitute) and water after contact with stools can reduce diarrhoeal disease by 35 percent or more.
- Eye and skin infections can be reduced with more frequent face and body washing.
- Improved water supply is generally associated with a 15 percent reduction in diarrhoea.

The most recent estimates of global costs to attain the water supply and sanitation MDG by Hutton and Bartram (2008) show that the spending required in developing countries on new coverage to meet the MDG target to 2015 is US$42 billion for water and US$142 billion for sanitation, a combined annual equivalent of US$18 billion. The cost of maintaining existing services totals an additional US$322 billion for water supply and US$216 billion for sanitation, a combined annual equivalent of US$54 billion. Spending for new coverage is largely rural (64%), while for maintaining existing coverage it is largely urban (73%). Additional programme costs, incurred administratively outside the point of delivery of interventions, of between 10% and 30% are required for effective implementation.

Evans *et al.* (2004) report that 1.47 billion people (20% of the world's population in 2015) would benefit if the sanitation target is met and this would increase to 2.16 billion if water and sanitation are both addressed. Total economic benefits of reaching the sanitation target may be of the order of US$63 billion annually (Table 15). This increases to US$225 billion annually if 100% access could be achieved. The bulk of the economic value of these benefits is associated with time savings. Health benefits from improvements in water and sanitation are summarized in Box 5 and Table 16 and examples of non-health benefits to the consumer as well as farmer and industry worker are summarised in Table 17.

Table 14: Examples of annual cost estimates to meet the water and sanitation MDG Target through 2015

Annual costs (billion USD)	Comments	Sources
1.6	for most basic level of sanitation	UN Millennium Project, 2005
3.1	for simplest possible approaches	Evans *et al.*, 2004
10.0	to supply low-cost services to those not currently supplied	Toubkiss, 2006
11.3	for new coverage	Hutton and Haller, 2004
13.0	for water supply	Winpenny, 2003 - Camdessus Commission
15-20	to provide higher level of services and maintain current level of service to those already supplied	Toubkiss, 2006
16.6	for sub-Saharan Africa, East Asia & Pacific, South Asia	Hutton, Haller and Bartram, 2006
17.0	for sanitation - these figures do not take into account wide sector management cost, as well as operation, and maintenance cost	Winpenny, 2003 - Camdessus Commission
18.0	for new coverage	Hutton and Bartram, 2008
30.0	for both water and sanitation	WEHAB, 2002
54.0	for maintaining existing services	Hutton and Bartram, 2008
80.0	for most expensive technologies	Evans *et al.*, 2004
200	assuming that conventional technologies are used	Bos *et al.*, 2005

Table 15: Examples of economic benefits of meeting the water and sanitation MDG Target

Annual economic benefits (billion USD)	Comments	Sources
38.0	total economic benefits for meeting the water and sanitation targets estimated based on six coverage scenarios sorted by world region	Hutton, Haller and Bartram, 2006
63.0	total economic benefits of reaching the sanitation target; this exceeds US$ 225 billion annually if 100% access could be achieved; the bulk of the economic value of these benefits is associated with time savings	Evans *et al.*, 2004
7.3	in terms of health-related costs avoided; the annual global value of adult working days gained as a result of less illness would be almost US$ 750 million	Bartram *et al.*, 2005

Table 16: Examples of health benefits from water and sanitation improvements

Health benefits	Source
155 million cases of diarrhea prevented, increasing to 546 million cases prevented for the water and sanitation MDG, and 903 million for universal access to water supply and sanitation	Hutton, Haller and Bartram, 2007
global reduction of diarrhoeal episodes of around 10%	Bartram *et al.*, 2005
improving access to water and better hygiene can reduce trachoma morbidity by 27%	UN World Water Development Report, 2003; Bartram *et al.*, 2005
basic sanitation can reduce schistosomiasis by up to 77%	UN World Water Development Report, 2003; Bartram *et al.*, 2005
safe drinking water and basic sanitation combined with hygiene can reduce morbidity from ascariasis by 29%	UN World Water Development Report, 2003; Bartram *et al.*, 2005

Table 17: Examples of non-health benefits of meeting the water and sanitation Target of MDG 7

(Adapted from Hutton and Haller, 2004)

Non-health benefits	Beneficiary
• More efficient managed water resources and effects on vector bionomics	Health sector/ Patient
• Time savings related to water collection or accessing sanitary facilities; • Labour saving devices in household; • Switch away from more expensive water sources; • Property value rise; • Leisure activities and non-use value	Consumer
• Benefits to agriculture and industry of improved water supply, more efficient management of water resources – time saving or income generating technologies and land use changes	Agricultural and industrial sector

6.2. Shortcomings of existing costing estimates

According to Evans *et al.* (2004), WHO estimates are the most sophisticated currently available as they take into account existing levels of services and incremental improvements. There are, however, several shortcomings of existing costing estimates. This is as a result of methodological weaknesses or simplified assumptions that make the cost estimates difficult to interpret (Hutton and Bartram, 2008). WHO estimates fair assumptions about recurrent costs, but uses nationally-provided unit investment costs. Also cost figures might underestimate the total requirements as they do not take into account wider sector management costs as well as operations and maintenance costs of existing stocks (Mehta *et al.*, 2005).

Estimates of the costs of reaching the 2015 target vary widely due to different approaches as well as the weak information base from which all estimates must be made (Evans *et al.*, 2004). The greatest weakness is that we don't have adequate cost data derived from the local level. This also means that the whole element of local economic trade-offs which is how communities work in reality is not touched upon in the global economic assessments. Most studies have ignored the costs of maintaining existing coverage levels - i.e. the costs of operating, maintaining, monitoring and replacing existing infrastructure and facilities (Hutton and Bartram, 2008). Moreover, the existing costing studies are principally based on economic capital.

There is yet to be a study that would estimate the cost of attaining the MDG Target for water and sanitation including the value of reused nitrogen, phosphorus and potassium, the costs of which have increased recently by three to five-fold because of the dependency on fossil fuels and depleting of phosphate reserves in the United States. A comprehensive study of costs of NPK fertilizer products including the mark-up from factory to farmer, and the quantity of fertilizer applied by farmers in different regions of the world is greatly needed.

We still lack, in general, cost estimates from the user perspective, be it the (non-backyard) farmer who is interested in urine from public toilets or a smallholder farmer relying on nutrients from neighbourhood toilets. In addition there are the transport costs of moving sanitation products to the farmers. The whole question of willingness-to-pay and spending priorities for poor communities also require more investigation. There is a great need for more precise and detailed cost estimates of sanitation options. In addition, little is written about time lost due to improper or dysfunctional toilet facilities, as use of toilet or personal hygiene are rarely included in questionnaires about time use. Finally, there is need to obtain data on how diarrhoeal disease burden varies between different income groups.

6.3. Cost comparisons

Black and Fawcett (2008) stress that conventional waterborne sewerage is not an affordable way of dealing with the sanitary crisis in non-industrialized countries and low-income communities, and it is impossible to picture a time when it could become affordable in large parts of the developing world still characterized in this way. On the other hand, composting toilets, once deemed an extreme solution or for areas without access to municipal sanitation, have evolved into sophisticated apparati that many now prefer to conventional toilets (Herro, 2006). Table 18 provides some cost comparisons.

Achieving the MDG target for sanitation using conventional approaches would provide multiple positive spin-offs (Hutton and Haller, 2004). These result in relatively high benefit to cost ratios averaging 5.5 for all targeted regions (8.9 for sub-Saharan Africa) only examining the health and social benefits, letting alone the environmental and reuse aspects.

However, to reach this and go even further, within the MDG time frame, innovation focusing on sustainability is badly needed within the water and sanitation sector and should be central to the strategy as laid out by the MDG Task Force on Water and Sanitation (UN Millennium Project, 2005). Choosing more advanced types of technologies such as provision of regulated in-house piped water would lead to massive overall health gains, but it is also the most expensive intervention (Hutton and Haller, 2004). Conventional approaches to sanitation, even if successfully applied to the MDG target populations, do not address the present situation for those today that receive poor standard sanitation services. This is a major source of health and environmental problems. Also, the existing conventional approaches to sanitation might not always be appropriate to the context of the MDG target population for a variety of reasons. Thus there is a need for a new look at sanitation using some principles of sustainability. Often ecological systems are perceived as being more expensive than conventional ones. But a cost comparison (Table 19) shows that both initial capital costs and those for operations and management are in fact very competitive. This is true for both composting systems like the Arborloo and the urine diverting dry vault systems.

Figure 29 illustrates the broad range of the unit capital costs for various sanitation options. Here it can be seen that the wet and dry off-site systems to the left require the highest investments while the decentralized condominial sewerage systems and on-site wet and dry systems require much lower unit investments.

Table 18: Some cost comparisons between piped sewerage systems and other systems

Sources	Estimates
WHO & UNICEF, 2000	average per capita construction costs of sewer connections or septic tanks in developing countries are higher (US$ 115-160) compared than pour flush toilets (US$ 50-90) and simple or VIP latrines (US$ 39-60).
Hutton & Haller, 2004	annual per capita running cost for sewer systems – US$ 5-13.4, US$ 9-12.4 for septic tanks, and US$ 4-6.4 for pit latrines.
Shilton & Walmsley, 2005	cost to treat 100 cubic meters of sewage may be as low as US$ 2-10 (for mechanical treatment, flocculation and others) or as high as US$ 60-100 (for activated carbon absorption or ion exchange).
Von Sperling & Chernicharo, 2005	pond treatment systems can provide good performance for relatively low investment (US$ 15-40 per person) and recurrent costs (US$ 0.8-3.5 per person and year.
Fonseca, 2007	annual per capita cost for sewerage connection lies between US$ 24-260. On-site solutions tend to be much more cost effective with annual per capita costs of US$ 11-54 for a simple pit latrine, US$ 10-172 for a VIP latrine but up to US$ 799 for a septic tank.

Table 19: Selected cost comparisons between ecological and conventional systems including capital investments and running costs

Sources	Comparison
Ecosan Club, 2003	compared expenses of flush toilets with a mechanical and vertical subsurface constructed wetland treatment with urine diversion dehydration toilets (UDDTs) and a horizontal subsurface constructed wetland system for a girls school in rural Uganda. Investment, reinvestment, and O&M costs were calculated during a 50 years time-frame with an annual interest rate of 8%. Results showed that the conventional system would be about **60% more expensive** than the ecological sanitation alternative, mainly due to the more compact wetland system and the additional expenditures on pumping required for the conventional system.
Holden *et al.*, 2004	found that urine diversion systems are more affordable compared to other contemporary solutions in South Africa. They have lower unit capital **(ZAR 1500)** and running **(0 ZAR/yr)** costs than VIP latrines **(ZAR 2000; 200 ZAR/yr)** or a waterborne system **(ZAR 10 000; 1200 ZAR/yr).**
Baten & Mels, 2004	compared the costs of advanced small wastewater treatment plants with a source separation system in rural regions of the Netherlands – the investment costs of the ecological sanitation system were estimated to be 15% higher, but the running costs were 25% less.
Mayunbelo, 2006	gives a comprehensive review on costs of VIPs and UDDTs for the entire population of Lusaka, Zambia's capital city. He found that both capital (**43,3 million Euros)** and running costs (**2.8 million Euros/yr**) for UDDTs would be lower than VIP systems (**47.7 million** and **3,1 million Euros/yr** respectively. The total costs for 10 years of operation in NPV would be 59 million Euros for UDDTs compared to 65 million Euros for VIPs.

Sources	Comparison
Abaire & Shane, 2007	report a substantial lower construction cost of an Arborloo in rural Ethiopia (US$ 5-12) compared with simple latrines (US$ 33-46) and VIPs (US$ 70-90).
Olbrisch, 2006	in Durban (South Africa), construction costs were in the range of **331-398 Euros** for a VIP, **398-597 Euros** for urine separation toilets (depending on the size) and about **928 Euros** for septic tank systems. Emptying septic tanks is done every 5-8 years and costs **133 Euros** while annual costs of maintaining urine separation toilets are **3.30-4.65 Euros.**
De Silva, 2007	compared unit costs for UDDTs and VIPs in Accra (Ghana) and found that investment costs per capita were **39 Euros** for UDDTs and **27 Euros** for VIPs while annual operation cost (including transport, treatment and sales of by-product) were deemed similar (**2.2** and **2.1 Euros**, respectively)

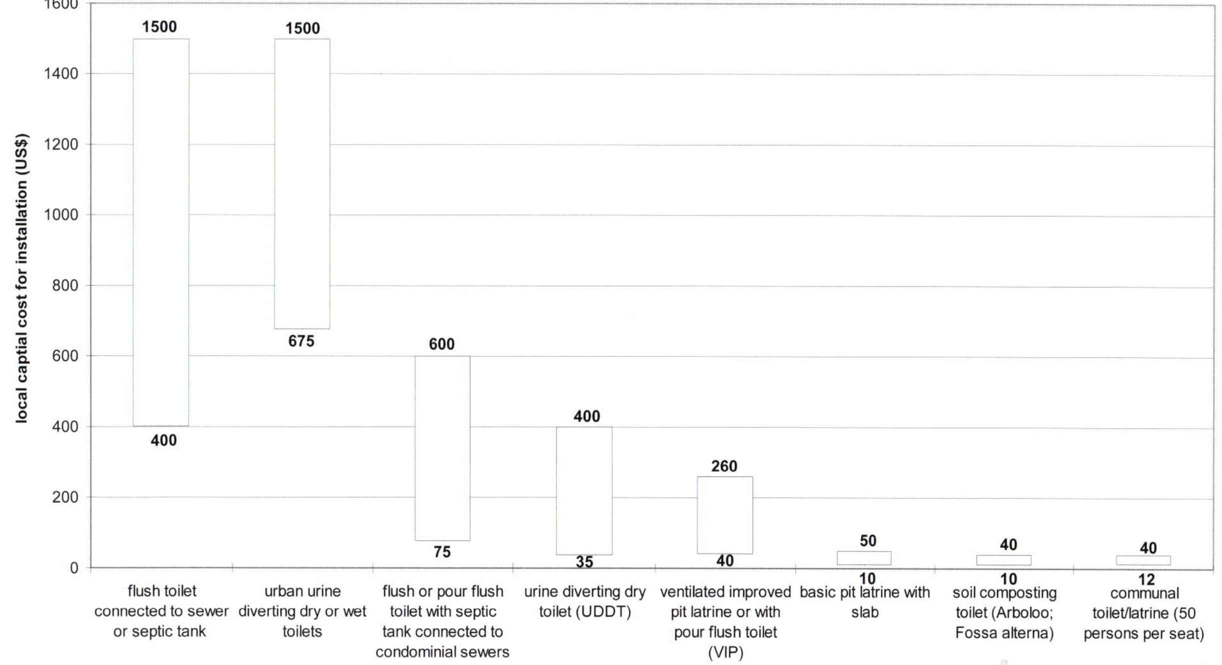

(Sources: UNDP, 2006; Satterthwaite and McGranahan, 2007; Water and Sanitation Fund of Namibia, 2008; UNICEF-SEI India, 2008; WESnet India, 2008; SEI, 2005)

Figure 29: General cost ladder for various sanitation options

7. Achieving Sustainable Sanitation

To provide the world with more sustainable sanitation solutions will require enormous efforts in the areas of capacity development, finance schemes, institutional collaborations at local, national, regional and international levels, improved planning including stakeholders, awareness raising, and large efforts in societal and technical outreach. The following section deals with the challenges of placing sanitation higher up on the agenda. It also lists many of the steps required in order to ensure that at least some of the new and retrofit sanitation programmes contain elements of sustainability.

7.1. Key drivers

WaterAid (2008b) recently published a call for action titled "Giving Sanitation the Green Light" focusing on sub-Saharan Africa. It mentions several of the key issues that must be addressed in order to promote sanitation including: leadership on the part of African governments, political neglect which is the root cause of the lack of progress and the low level of financing both from international and national sources. The call for action also lists several key questions of concern namely:

- is there a national sanitation policy;
- are national targets in line with the MDG target;
- what weighting is given to sanitation in the PRSP;
- is there a sector investment plan;
- is there a single body to coordinate action;
- are donors coordinating their support to sanitation;
- is there sufficient budget allocation to meet targets;
- is there a single budget line for sanitation;
- is there a performance monitoring mechanism?

Taking a practical focus, Sanitation 21 (IWA, 2006) provides a good overview of key drivers or parameters that are central to sanitation promotion and development. At the city and national levels these include:

- equity and sanitation access
- economic development
- reduction of costs
- water resources management

At the household level the concerns surrounding sanitation include:

- status, cleanliness and convenience
- safety and security
- income/employment generation
- food and water security
- land tenure status and legislation
- health and the environment

The challenge is to link these tangible needs to institutional arrangements at the local, and national levels in order to ensure that the needs can be met. Marrying the parameters cited by WaterAid with the concerns described in Sanitation 21 is the sanitation challenge in a nutshell.

7.2. Institutional players

There is a large number of organisations involved in different ways with sanitation questions. Still the enormous challenges described in this paper exist. At the international level, several organisations have explored and are developing strategies for this to happen. These include the World Bank Water and Sanitation Programme; UNICEF; WHO; UN-Habitat; International Federation of the Red Cross, UNDP; UNEP; UN Water; the United Nations Interagency Gender and Water Task Force; the United Nations Secretary General's Advisory Board on Water & Sanitation; the African Development Bank African Water Facility; the Asian Development Bank Water Financing Programme; the UNESCO-Institute for Water Education; the International Water and Sanitation Center (IRC); the International Water Management Institute; the Global Water Partnership; Plan International; KfW Bank Group; GTZ; Sida; IDRC; DGIS; SDC; Water, Engineering and Development Centre (WEDC) at Loughborough University; Technical University Hamburg-Harburg; Norwegian University of Life Sciences; Swedish University of Agricultural Sciences; Swedish Institute for Infectious Disease Control; London School of Hygiene & Tropical Medicine; WASTE; EAWAG-Sandec; International Training Network; Catholic Relief Services; Water Aid; the Gender and Water Alliance; the Water Supply and Sanitation Collaborative Council (WSSCC); AKVO (Open Source for Water and Sanitation); CARE; the World Toilet Organisation; Stockholm Environment Institute; and the Sustainable Sanitation Alliance (SuSanA).

Regionally and nationally, there are several organisations that have been instrumental in putting sanitation onto the development agenda. A sampling includes CINARA in Colombia, SARAR in Mexico, CREPA in West Africa, WaterNet and NETWAS in Africa, Cap-Net in various regions of the world, CSE, SCOPE and Sulabh in India, IWSD in Zimbabwe, Water Resources Commission in South Africa, Practical Action in South Asia and East Africa, CAPS and PEN in the Philippines, WECF in East Europe and Central Asia, MAMA-86 in the Ukraine, and Hesperian Foundation and Ecowaters in USA.

The year 2008 was the International Year of Sanitation (IYS) and this provided a global focus to the overall problems and challenges being tackled. The official website of the IYS 2008 (esa.un.org/iys/) provides a broad overview of what many of these organisations are involved with and also includes many of the key publications produced thus far. The regional conferences organised by the Water and Sanitation Programme, the annual World Water Week organised by Stockholm International Water Institute and the World Water Forum (WWF) organised every third year by the World Water Council have been also instrumental. The UN World Water Development Report Three to be released at the WWF in Istanbul in 2009 will attempt to summarize again the current status of knowledge worldwide.

With all this capacity it may be therefore surprising to observe the enormous gap between what is being written or said and what is actually happening on the ground. The gap is significant and has been generally attributed to a general lack of political leadership and commitment particularly at the national level of governance around the world. The cause for this lies essentially in the fact that there is little public outcry or dialogue on the sanitation crisis (Rosemarin, 2007) resulting in little demand for political action. In a way the whole world needs a dose of CLTS (Community-Led Total Sanitation) not just to stop the open defecation by over a billion people, but to encourage people to face the fact that the “flush and forget” or “hide and forget” mentality is having serious short- and long-term consequences on human health, the environment and natural resources. It is therefore for good reason that the MDG target for sanitation will not be met by 2015.

7.3. Capacity development

Capacity development is understood as a process of unleashing, strengthening, creating and maintaining capacity over time. It applies to individuals, organizations and institutions. Capacity development is more than awareness of technical subjects and general organizational principles and it cannot be imported, but must be led from within the country itself. Capacity development within sustainable sanitation requires more than arranging occasional workshops and training courses. It is a dynamic process where learning links up with live experience to improve outputs, processes and products. Sufficient time and resources are necessary components to connect acquired capacity with action. Capacity development for sanitation has proven successful in different local and national contexts where the cultural understanding and ownership have been sufficiently incorporated (OECD, 2006).

The lack of capacity and the lack of demand from users prevent sanitation from entering into the local development agenda and the importance of sanitation is only usually recognized in times of emergencies. In parallel some responsibilities and services are delegated to private operators such as those constructing household latrines, emptying latrine pits and septic tanks, or managing the day-to-day operation of public latrines. Increased understanding is needed within local government about the range of forms of interaction between state and non-state providers of sanitation services concerning disposal, treatment and reuse for identification of which are most appropriate in a given situation, and how they can be managed and regulated.

Lack of capacity is a major constraint in achieving the MDG target because of:

- Limited absorptive capacity, i.e. inability to make use of available resources;
- Poor service delivery and performance;
- Limited transfer of knowledge;
- Construction of infrastructures without consultation with end-users

Clear goals have been defined and knowledge exists on how sanitation can improve socio-economic conditions and livelihoods of the un-served poor but reality reveals that political actions remain inadequate. Furthermore, there is lack of capacity and resources. Weak institutions and lack of political will for sanitation make it difficult to enforce and follow the existing legislation also making it difficult to introduce, implement and scale up sustainable sanitation and other innovative solutions/approaches.

Strategies to address capacity needs in sustainable sanitation

It is necessary to develop, in an innovative manner, strategies which enable capacity development to take place in regions with low sanitation coverage. Examples of some approaches are as follow:

- A stepwise alignment of strategies for planning, implementation and monitoring performance within a sector-wide approach can help to identify where the key capacity gaps are, and to find the most effective methods to address them;
- Driving forces toward sustainable sanitation vary between different regions and cultures, thus capacity development should be based first on the local problems as a framework. Experiences from successful projects need to be continuously documented, analyzed, adapted and replicated in new settings;
- Strategic partnership between small-scale providers, knowledge institutions and organizations that are actively providing training in sustainable sanitation are necessary to stimulate the market and advance quality assurance;
- Gender must be observed in the planning of capacity development and during transmission of essential messages to users. Women and men often have different possibilities to participate in training opportunities and may have different needs regarding content;
- Knowledge on how sanitation affects women, men and children differently needs to be taken in to consideration when planning, designing, constructing and maintaining sanitary facilities. Also the fact that the poor and vulnerable people that have little access to sanitation could be highlighted more. Since different target groups need to be identified and reached, training needs to be tailored to correspond to the background and aspirations of the new groups;
- Sanitation needs to be discussed regularly in the media, and information about champions in sustainable sanitation needs to be highlighted by the media;
- The language barrier needs to be considered. The availability of information across the language divide is very important.

Tools available

The efforts from the global sustainable sanitation community to strategically accelerate capacity development during recent years has been impressive and has resulted in a variety of methods which are outlined in Table 20 and sources listed in Box 6.

Box 6: List of websites and electronic newsletters including material on capacity development

- www.akvo.org
- www.caps.ph (CAPS Notes is the quarterly newsletter of the Centre for Advanced Philippine Studies and provides information and updates about urban waste management)
- www.ecosranres.org
- www.esa.un.org/iys/output/Documents_list.asp
- www.gtz.de/ecosan (includes quarterly electronic newsletter in English, German, Spanish, French and Chinese)
- www.genderandwaterorg.com
- www.hesperian.org (Hesperian publications are designed so that people with little formal education can understand, apply and share health information)
- www.irc.nl/page/116 (includes "Source Water and Sanitation News Service": weekly news in English, French and Spanish with an emphasis on rural and peri-urban areas in developing countries.)
- www.library.eawag-empa.ch
- www.wrc.org.za/publications_reports.htm
- www.unwater.unu.edu/article/698
- www.zer0-m.org
- www.wsp.org/index
- www.susana.org

Table 20: Tools for capacity development

Item	Comments
Reference centres for knowledge transfer	Resource centres positioned locally and globally focus on capacitating the sector by supporting the development of pro-poor policies for the sustainable implementation of water, sanitation, health and hygiene.
Regional knowledge nodes	To establish regional knowledge nodes in recognized institutions, networks and organizations specialized in sustainable sanitation that act as knowledge brookers in a region is a way of obtaining best regional knowledge and experience.
Courses and research groups	Universities in various locations are offering courses on sustainable sanitation. The ongoing curricula development has led to more flexible courses using innovative methods where e-learning is combined with face to face courses.
E-learning courses	With regard to capacity development of mid-career professionals in developing countries, online courses are a promising alternative (or addition) to face-to-face courses. This type of target audience is well versed with using computers and the internet, and is therefore a suitable target group for online courses.
Learning alliances	The learning alliance is a (i) mean for knowledge outreach and learning within the alliance, (ii) way to make sure that research carried out is relevant and that the results reach the stakeholders, (iii) a mean to upscale innovation within the sector.

Item	Comments
Networks	Knowledge sharing across networks is an efficient mechanism to make international knowledge accessible while the local network then adapts the information to fit local conditions. One strength of network is the ability of assemble a critical mass of expertise and resources.
Educational movies and video clips	The major strength of moving image media is its closeness to human perception and thus easy to comprehend by audiences with wide range of cultural and educational background. Several short educational movies on sustainable sanitation have been produced on different continents.
Street drama	This is mainly for people without access to written material or media. It involves a dramatized open air performance in streets, fields, or open spaces and not only provides entertainment but also arouses social consciousness and encourages the audience to act. This tool has been used very successfully for introduction and creating demand for more sustainable sanitation in countries in sub-Saharan Africa.
Internet discussion fora	This is mainly for global discussion and information sharing on knowledge development. There are now a number of discussion groups on various aspects of sustainable sanitation in several languages.
Technical Enquiry Service	The Technical Enquiry Service is able to call on the expertise of several hundred professionals in technical, economic, and sociological disciplines to help formulate the answers to enquiries. The Technical Enquiry Service always tries to supply information of direct relevance to the individual enquirer's circumstances and will take into account the non-technical factors that might have a bearing on the use of the technology.
Publications and websites	Electronic copies of publications on sustainable sanitation can be found free for download from various open sources and libraries.
SuSanA website and DVD with resource material	The SuSanA website and DVD contains ample capacity development material on the topic of sustainable sanitation and is grouped by topic, working group theme, by language, by country and by cross-cutting theme.

7.4. SuSanA

The Sustainable Sanitation Alliance (SuSanA) was formed in January 2007 in order to provide a global focus on sanitation and to provide capacity and knowledge within the sector. As of August 2008 it had 93 participating organisations (Figure 30). These vary from NGOs and expert consultancies to research organisations, UN agencies, bilateral donors and development banks. SuSanA formed 12 thematic working groups as follows linking sustainable sanitation and:

- capacity development
- costs & economics
- renewable energy/groundwater/
- climate change
- technology options/hygiene/health
- food security
- cities & planning
- community & rural sanitation
- emergency & reconstruction
- sanitation as a business
- public awareness & marketing
- operations & maintenance

The goals and objectives of SuSanA are to raise awareness about the need for sustainability within the sanitation sector and sanitation systems, to promote such systems at a large scale, to identify the key role that sanitation plays in human development and the MDGs, to emphasise the importance of stakeholder participation in sanitation and hygiene planning and programme execution and help change the general attitudes about sanitation from a disposal-oriented one to a more reuse-oriented one.

Fact sheets are being produced by each of the thematic working groups and these function as summaries of the latest state of knowledge. In addition a resource DVD has been produced and a website has been launched containing all the knowledge products.

SuSanA has held several meetings hosted by participating partners in Europe, North America, Africa and Asia during its two first years of operation. At present its secretariat is housed at GTZ in Germany with administrative support from SEI/EcoSanRes in Stockholm.

7.5. Financing sanitation initiatives

As described in this paper, the direct and indirect positive effects of sanitation investment average about 10-fold when taking into account health and societal gains. Adding the environmental gains to this total should therefore without any doubt make sanitation investments a foundation to poverty elimination and improvement in livelihoods and well-being. The challenges arise, however, in exactly how these investments are to be made and whether the onus is on the government or the individual consumer to foot the bill. The sector is characterised by subsidy financing both in the North and the South. This approach has divided the sanitation world into an urban and rural one, where the former is highly subsidised and the latter left on its own with minimal subsidies. The gap between water supply and sanitation was reviewed in this paper but an even more obvious gap exists between the urban centres and peri-urban zones which in many developing countries are classed as informal settlements. It is in these informal settlements that sanitation services are most wanting. Planning sanitation systems in these three distinct zones therefore requires rather different approaches. But at the same time there are elements they have in common.

Figure 30: SuSanA objectives, activities and partners

In 2003, the Camdessus Panel (Winpenny, 2003) warned that the MDGs would not be achieved unless annual investments in water supply and sanitation services in developing countries are doubled from the 2003 level (US$15 billion per year to US$30 billion per year). The Camdessus Panel provided a series of 87 proposals grouped under the following categories:

- Enhancing sub-sovereigns' access to finance (8 proposals)
- Decentralisation of water services and fiscal relationships (5)
- Promoting local capital markets (3)
- Promoting private sector participation (11)
- Adapting financial policies and instruments to the needs of the water sector (14)
- Capacity building and technical assistance (10)
- Increasing aid to the water sector (4)
- Improving aid effectiveness (13)
- Improving water services efficiency and sustainability (5)
- Water policy and water sector framework (10)
- Monitoring and reporting (5)

The report of the Camdessus Panel was endorsed by G8, the World Bank, AsDB and others. The work of the panel was to be followed up by the Gurria Task Force (www.financingwaterforall.org) by 2006 at the 4th World Water Forum in Mexico, but it provided no specific report on progress, only providing some case histories from a few countries illustrating the demand and supply side of financing water projects (none of which were from sub-Saharan Africa).

The advice of the Camdessus Panel has unfortunately had little impact. The sector has only experienced marginal finance increases since 2003 (Shah, 2007) and this has resulted in the growing view and need to provide local level financing from local governments. This combined with a strengthened political view that water and sanitation services are best dealt with by local and national governments may in fact be the single most important development from all these deliberations.

Local financing, however, requires necessary building blocks in order to generate necessary demand-driven and sustainable sanitation:

- Demand needs to be based on awareness and knowledge about the advantages of hygiene and sanitation
- Supply includes the capacity to deliver products that work, that are affordable and acceptable to the users and that can be maintained
- The sanitation system chosen needs to protect health and the environment and is not a source of pathogenic or chemical contamination of ground and surface waters

- Integrated policies need to drive the process bedded in the fact that sanitation pays for itself with high benefits to costs ratios
- Financing has to be both predictable and reliable over the long term and not be based on external subsidies

The long-term predictability and reliability of the financial sources are paramount:

- In developing countries, the lack of predictability and reliability of subsidies and external financing is a principle constraint
- This is especially true if O&M (operations and maintenance) is based on subsidies that are not dependable
- Government priorities change (reflecting the lack of strong national policies)
- Donor funds are just as unpredictable
- Dependence on donor grants lowers the chances of achieving long-term financial sustainability
- Loaned funds are subject to competing local priorities
- A private sector utility can be financially stable if it is creditworthy but requires local governance to ensure reliable performance

The sanitation gap across the world is directly related to GDP or ability to pay both at the government and consumer levels as illustrated in Figure 31. The poorest countries have the largest backlog in terms of installations required.

An attempt was made to determine what percentage of the GDP is required to invest in sanitation systems (SEI, 2005). As can be seen in Figure 32 the total regional annual costs for ecosan between 2003 and 2015 would require less than 0.2% of the domestic GDP for each UN region. It is only in East Asia and Sub-Saharan Africa where the cost edges over 0.2%. In Eurasia, Occania and West Asia it would cost 0.1% or less. These levels when compared to other expenses such as health or military are therefore well within the domestic limits of affordability. That is, there should be sufficient financing available at the regional level to cover the MDG requirement. Global public expenditures on health care were 5.8% of the GDP and those for military were 2.4% (World Bank Group).

Of course the key in making sanitation sustainable is the operations and maintenance in order to ensure the facility functions properly over its designed lifespan. Annual improvement costs were calculated for various sanitation schemes including the price of materials and equipment of the latrine, labour and equipment for installation, and maintenance costs over the lifetime of the solution, discounted at 3% (Table 21) (Hutton *et al.* 2006).

These annual improvement costs per person can be covered within the 1% GDP benchmark (Table 22) but very often are neglected in annual investments.

Table 21: Annual improvement costs per person reached, nominal 2000 US$

(Source: Hutton et al. 2006)

Intervention	Africa	Asia	LAC
Septic tank	$9.75	$9.10	$12.39
Ventilated improved pit latrine	$6.21	$5.70	$5.84
Small pit latrine	$4.88	$3.92	$6.44
Household sewer connection plus partial sewer treatment (hardware and software)	$10.03	$11.95	$13.38
Household sewer connection plus partial sewer treatment (software only)	$4.84	$5.28	$6.46

Table 22: GDP, household size and calculation of 1% GDP per household for the 9 MDG regions

(Source: SEI, 2005)

UN Region	GDP/cap (2003 USD)	Household size	GDP per hh	1% of GDP per hh
East Asia	$1,486	2.73	$4,063	$41
Eurasia	$2,016	2.25	$4,536	$45
Latin America & Caribbean	$3,393	3.53	$11,967	$120
North Africa	$1,582	4.80	$7,595	$76
Oceania	$1,554	4.78	$7,433	$74
South-East Asia	$1,364	3.81	$5,200	$52
Southern Asia	$599	5.08	$3,046	$30
Sub-Saharan Africa	$625	4.29	$2,685	$27
West Asia	$4,906	4.95	$24,281	$243

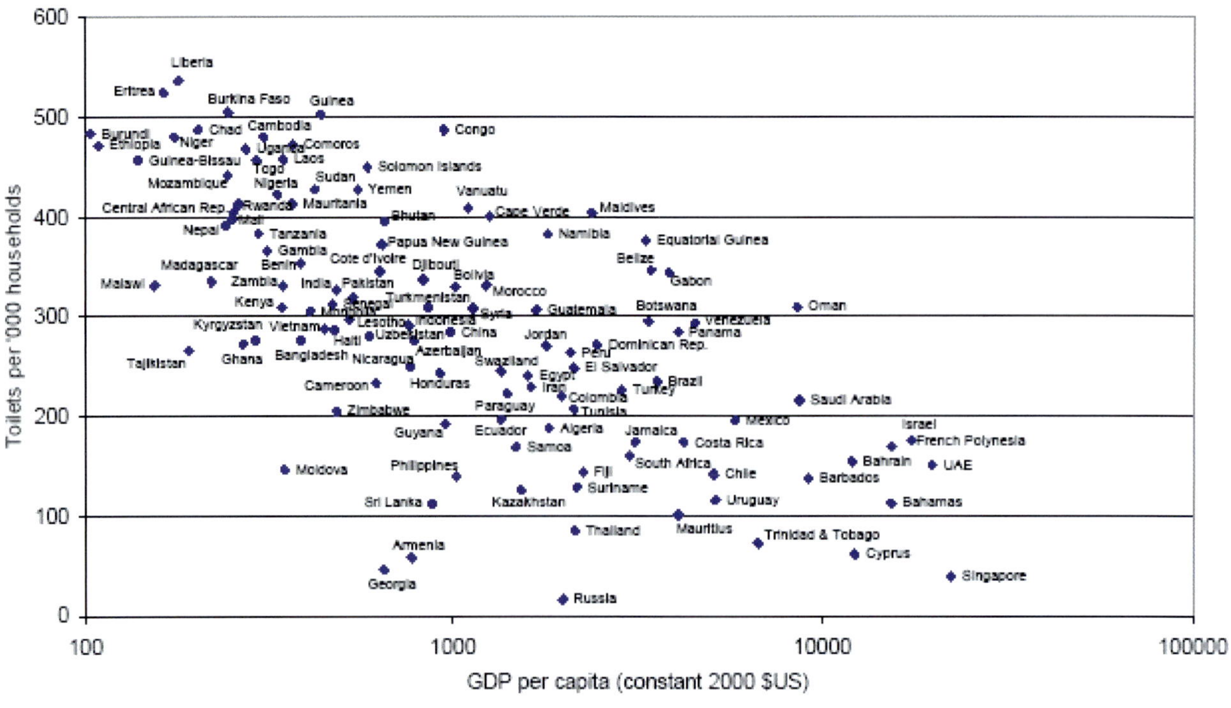

(Source: SEI, 2005)

Figure 31: National sanitation requirements as number of toilets per thousand households to reach MDG 7 Target 10 in relation to GDP per capita

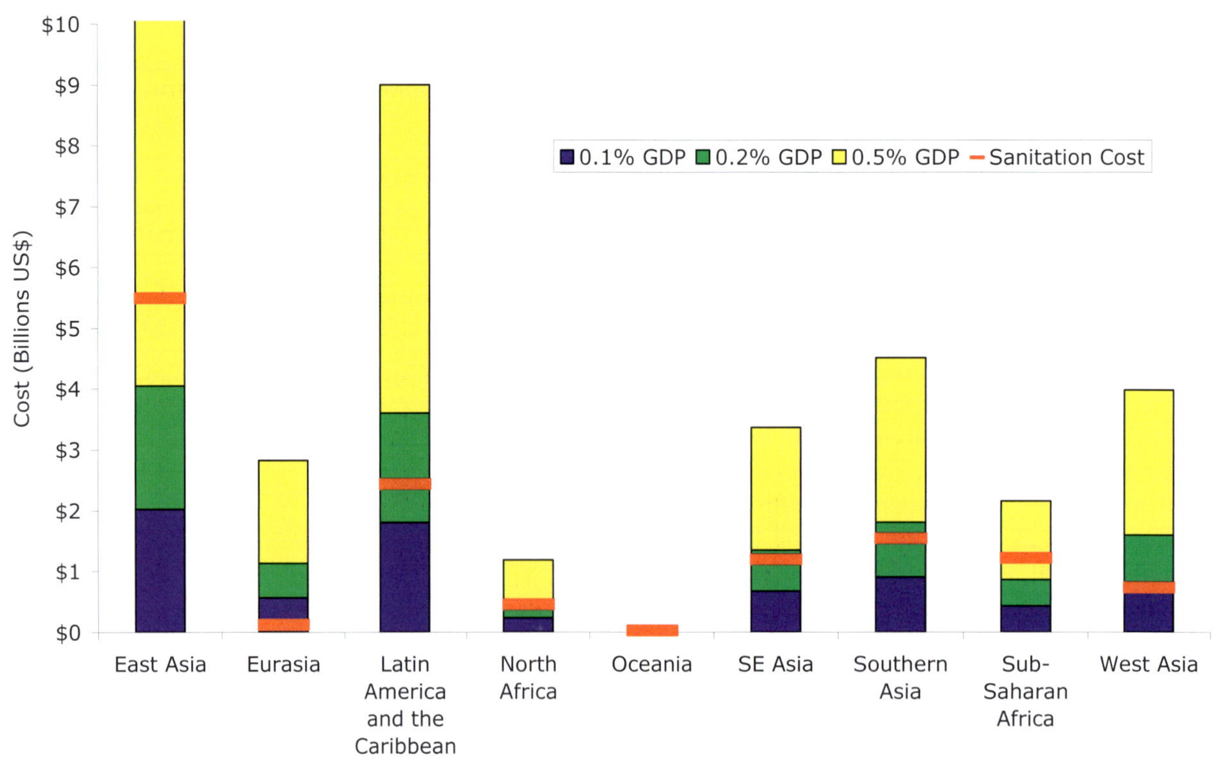

Sanitation cost (red horizontal bars) as annual expenditures (Y-axis). The blue, green and yellow bars are the GDP expenditure levels of the regions also identified as 0.1, 0.2 and 0.5% of the GDP respectively. (Source: SEI, 2005)

Figure 32: Cost for ecological sanitation investments between the period 2003 and 2015 to meet the MDG sanitation target

There is also a need to introduce user charges and commercial financing terms to build in financial resilience:

- Household and user needs must be listened to and given the highest priority
- If the demand is well-defined and the sanitation systems are appropriate and affordable for the community, commercial terms are possible
- Installations can be financed using credit loans
- O&M can be financed based on user fees, taxes and surcharges

Also urban systems have to be managed differently from rural ones:

- Urban sanitation requires
  - decentralised or centralised utility-based systems
  - a service-based approach
  - loans for infrastructure and user tariffs
- Rural sanitation requires
  - onsite household-based systems
  - strong linkage to rural development, land tenure, agriculture extension and health services
  - microfinancing loans
- Both require financial grants for awareness raising, stakeholder training and demonstration in order to leverage local market interest

Traditionally, sanitation financing schemes were concentrated to core urban area projects since these were considered most credit-trustworthy. With the introduction of microcredit schemes, however, rural projects have also received more attention.

There are several examples of innovative finance schemes some of which are listed here:

- UN Capital Development Fund – microfinance of inclusive financial sectors with permanence
- Microcredit schemes
  - Consultative Group to Assist the Poor (includes 34 donor members - multilaterals, development banks, bilaterals, foundations)
  - Grameen Bank (Bangladesh) - 50 million borrowers
  - Association for Social Advancement (Bangladesh) – 6 million borrowers
  - Foundation for International Community Assistance Village Banking Campaign to set up 100,000 village banks (Latin America, Central Asia, Africa) – currently 600,000 clients
- Cities Alliance (World Bank)
  - Specialises in upgrading urban slum areas
  - Supports City Development Strategies
  - Includes the Community-Led Infrastructure Financing Facility
- Philippine Water Revolving Fund – leveraging public resources with private funds affordable to local utilities

The US Revolving Fund Initiative which specialises in providing loans to sound projects has produced a wealth of experience summarized as follows:

- Construction grants are not sustainable over the long term and are thus not supported
- A functioning system to collect community user fees is a requirement for success
- Partnering with local funding sources, such as banks, non-profit groups, local governments, and state agencies can ensure the best financing for each community
- Legal mechanisms should be written broadly to allow both flexible and innovative financing by local entities and also to allow funding of projects not foreseen at the outset of the initiative
- Local agencies need to have effective management and regulatory procedures in place for successful project execution and financial stability

The concept of a "Global Bond Fund" can also be an option:

- Bonds are used to attract investment and generate money up front
- If the risk is low, bonds can be given high credit rating and become very attractive investment objects for financial institutions
- The Global Alliance for Vaccines and Immunization (GAVI) Fund first financed directly by the Gates Foundation was shifted to an alternative financial form where pledges from donor countries are sold as bonds on the market

A "World Sanitation Fund" could be proposed as well, where loans and small grants are provided to initiate sanitation projects:

- This would be an independent body with its own governance system to grade and approve projects
- The participating financial institutions would provide the predictability and reliability lacking in developing countries
- There would be grants for promotion, awareness raising, capacity building, training and demonstration programmes (to generate greater demand)

- And there would be larger loans for large-scale implementation

An example of innovative financing is the project in Orissa where the Dutch organisation WASTE together with the Indian micro-credit organisation BISWA, the Indian insurance company TATA-AIG and the Dutch financial group SNS-REAAL, have introduced low interest loans for sanitation for villagers (www.waste.nl/page/1546). The project is built on the assumption that sanitation improvement will directly improve health status and the effects are being monitored using health insurance data. The goal is to reach 500,000 households before 2012 and the project budget is 30 million Euros.

In summary:

- Demand for sanitation services still needs stimulation in developing countries
- Successful financing needs to be predictable and stable
- External financing needs to be used strategically if impacts are to remain long-term
- Self-help and community-led programmes are keystone to successful sanitation projects
- Grants may be used to kick-start the leverage of local capacities
- Loans may then be used to implement scaling up and may be financed by public and even public-private co-ventures
- Fees and tariffs may be used to finance operations and maintenance
- Sanitation can be financed using less than 1% of the GDP in most countries

7.6. Overall conclusions

In summary, this paper:

- provides an overview of the present status of improved sanitation in the developing world based on currently available surveys
- discusses the definition of improved sanitation and the need to begin introducing the elements of sustainability
- describes the linkages between the MDG 7 Target 10 on water and sanitation and the other MDGs including human health
- reviews the latest data for the various regions of the world on disease burden caused by lack of hygiene and sanitation
- describes the agriculture challenges and dependency on high priced fertilisers in relation to opportunities in exploiting "productive" sanitation
- describes the potential impacts of safe reuse of human excreta in agriculture, nutrition and food security especially for smallholder farmers
- reviews the criteria surrounding the introduction of sustainable sanitation, what technical options exist and what the relative costs would be compared to conventional approaches
- reviews the essential drivers behind increasing demand for sanitation
- reviews the various available planning tools for sanitation
- reviews the additional steps required including financial and institution arrangements in order to build resilience within the sanitation sector

This paper therefore concludes that:

- a curvilinear projection of the JMP sanitation coverage data from 2006 indicates that the MDG target could be met by 2019 as opposed to the linear projection estimate of 2037
- since the current UN definition of improved sanitation does not strictly take into account dysfunction and contamination of the environment, that the coverage data may be providing an unrealistic picture and in the future should incorporate more sensitive monitoring of management and sustainability
- if a more rigorous definition including sustainability criteria were to be used, the global sanitation crisis would be even larger than it is perceived to be today
- since the UN MDGs are generally not being tackled using sustainable systems as a unifying approach, that many of the gains and improvements made by 2015 may prove to be only short-term
- the health burden (e.g. diarrhoeal diseases and worm infestations) imposed by dysfunctional or non-existent hygiene and sanitation should be given as high a priority as other diseases like malaria, HIV/AIDS and tuberculosis
- informal settlements such as slums and peri-urban areas require special added attention in terms of provision of safe water and sanitation systems
- the introduction of sustainability criteria into the definition, planning and implementation of sanitation systems will have long-term positive impacts and make the investments even more cost-effective
- more appropriate, affordable and resilient sanitation systems are available than those currently being chosen and that professionals need to be better informed and trained about these
- even if the data show that sanitation pays for itself several times over in terms of improved health and livelihoods, and there are many ac-

tive institutional players involved, the lack of public dialogue and awareness prevents large strides in progress
- sanitation must be seen as an interplay between human behaviour (cultural attitudes and norms) and appropriate technologies requiring stakeholder and gender-sensitive involvement in the planning and implementation steps
- scaling up of sanitation in developing countries is hampered by inadequate planning systems, lack of public demand, inadequate local political leadership and unstable financing schemes
- financing of sanitation systems needs to be predictable and reliable and based on the local ability to pay and not entirely on external subsidies
- the costing out of sanitation systems must go beyond the initial capital expenditures and include operations and maintenance in order to provide resilience and long-term performance
- innovative financing schemes can be developed making use of micro-credit loans involving inclusive financial sectors that up to now have not been linked to sanitation
- more capacity building in the sector at the individual and institutional levels is needed in order to lift the sanitation sector into the era of sustainable development
- sanitation products such as water and nutrients should not be seen as waste products but valuable resources and that sanitation systems should be designed around possible reuse options
- introduction of the safe handling and reuse of the products from sanitation will have significant impacts on nutrition and food security in developing countries,
- in sub-Saharan Africa productive sanitation can significantly help substitute the use of chemical fertilisers.

8. References

Abaire, B. and Shane, L. 2007. Ecological Sanitation: Arborloo is Breaking Sanitation Challenges. Presentation of papers: World Toilet Summit 2007, New Delhi, India.

Bartram, J. 2008. Improving on Haves and Have-Nots. Nature 452(20): 283-284.

Bartram, J., Lewis, K., Lenton, R., Wright, A. 2005. Focusing on improved water and sanitation for health. Millennium Project. www.thelancet.com. 365: 810-812.

Baten, H. & Mels, A. 2004. Source-oriented sanitation in rural regions. In: Werner *et al.* 2004. Proceedings of the 2nd International Symposium on Ecological Sanitation in Lübeck, Germany. GTZ, Eschborn, Germany. Available at: http://www.gtz.de/de/dokumente/en-ecosan-symposium-luebeck-session-g-2004.pdf Accessed on the 9th February 2008.

Black, M. and Fawcett, B. 2008. The Last Taboo: Opening the Door on the Global Sanitation Crisis. Earthscan, UK and USA. 254p.

Borkowski, L. 2006. Environmental health, water. UNDP on the world's water challenge. Available at: http://thepumphandle.wordpress.com/2006/11/18/undp-on-the-worlds-water-challenge/. Accessed 19 February 2008.

Bos, A., Guzen, H., Hilderink, H, Moussa, M., De Ruyter, E. and Niessen, L. 2005. Health Benefits Versus Costs of Water Supply and Sanitation. Water and Health. IWA Publishing, London.

Brahmbhatt, M. and Christiaensen, L. 2008. Rising Food Prices in East Asia: Challenges and Policy Options. World Bank. 18p.

Capron, A., Dombrowicz, D. and Capron, M. 2004. Helminth Infections and Allergic Diseases: From the Th2 Paradigm to Regulatory Networks. Clinical Reviews in Allergy and Immunology. Humana Press Inc. 26(1): 25-35.

Chan, M. 1997. The Global Burden of Intestinal Nematode Infections – Fifty Years On. Parasitology Today, 13: 438-43.

Clarke, R. and King, J. 2004. The Atlas of Water. Earthscan, London. 127p.

CRDA (Christian Relief Development Association). 1999. The Role of NGOs in the Environmental Sanitation Sector, Addis Ababa.

De Silva, N. K. 2007. Multi-criteria analysis of options for urban sanitation and urban agriculture – Case study in Accra (Ghana) and in Lima (Peru). MSc Thesis. UNESCO-INE, Delft, the Netherlands. Available at: http://www2.gtz.de/Dokumente/oe44/ecosan/nl/en-kalyani-ihe-thesis-2007.pdf Accessed 17 April 2008.

Drangert, J.O. 1998. Fighting the Urine Blindness to Provide More Sanitation Options. Water South Africa 24 (2), April.

EAWAG. 2005. Household-Centred Environmental Sanitation: Implementing the Bellagio Principles in Urban Environmental Sanitation. Swiss Federal Institute of Aquatic Science and Technology. Duebendorf.

Ecosan Club. 2003. Sacred Heart Sisters Kalungu Girls Secondary School: Improvement of water and sanitation infrastructure. Available at: http://www.gtz.de/ecosan/download/study-ecosanres-ugandaschool03.pdf Accessed 3 March 2008.

Esrey, S.A., Potash, J.B., Roberts, L., Shiff, C.1991. Effects of improved water supply and sanitation on ascariasis, diarrhoea, dracunculiasis, hookworm infection, schistosomiasis and trachoma. Bulletin of the World Health Organisation. 69(5):609–621.

Evans, B., Hutton, G. and Haller, L. 2004. Closing the Sanitation Gap – The Case for Better Public Funding of Sanitation and Hygiene. Round table on sustainable development. Organisation for Economic Co-operation and Development (OECD). 25p.

FAO. 2006. The State of Food Insecurity in the World 2006. Eradicating World Hunger – Taking stock 10 years after the World Food Summit. FAO, Rome, 44p. Available at: ftp://ftp.fao.org/docrep/fao/009/a0750e/a0750e00.pdf Accessed 6 August 2008.

FAO. 2008a. Current World Fertilizer Trends and Outlook to 2011/12. Rome. 44p.

FAO. 2008b. Asia Pacfic Food Situation Update. FAO Regional Office for Asia and Pacific. 4p. Available at: http://www.fao.org/world/regional/rap/update.pdf. Accessed 6 August 2008.

FAO. 2008c. Food Price Indices, July 2008. Available at http://www.fao.org/world/regional/rap/update.pdf Accessed 6 August 2008.

Fonseca, C. 2007. Quantifying the Cost of Delivering Safe Water, Sanitation and Hygiene Services: An overview of cost ranges and trends. IRC report prepared for the Bill and Melinda Gates Foundation. Unpublished.

Friend, J. 1992. New directions in software for strategic choice. European Journal of Operational Research 61: 154-164.

Graff, G. 2007. Sulphuric Acid is Suddenly Scarce and Expensive. Purchase.Com Reed Business Information.

Gumbo, B. 2005. Short-cutting the Phosphorus Cycle in Urban Ecosystems. Dissertation, Delft University of Technology. Taylor Francis Group, London. 320p.

Hajer, M. A. and Wagenaar, H. 2003. Deliberative Policy Analysis: Understanding Governance in the Network Society. Cambridge University Press. 306p.

Herro, A. 2006. Composting Toilets Offer Solution to Water, Sanitation Problem. World Watch Institute – Vision for a Sustainable World. Available at: http://www.worldwatch.org/node/4751. Accessed 19 May 2008.

Holden, R. Terreblanche, R. & Muller, M. 2004. Factors which have influenced the acceptance of ecosan in South Africa and development of a marketing strategy In: Werner *et al.* 2004. Proceedings of the 2nd internationalsymposium on ecological sanitation in Lübeck, Germany. GTZ, Eschborn, Germany. Available at: http://www.gtz.de/de/dokumente/en-ecosan-symposium-luebeck-session-b-2004.pdf Accessed 12 May 2008.

Hutton, G. and Bartram, J., 2008. Global Costs of Attaining the Millennium Development Goal for Water Supply and Sanitation. Bulletin of the World Health Organization. 86(1): 13-19.

Hutton, G. and Haller, L. 2004. Evaluation of the Costs and Benefits of Water and Sanitation Improvements at the Global Level. WHO, Geneva. 87p.

Hutton, G., Haller, L. and Bartram, J. 2006. Economic and Health Effects of Increasing Coverage of Low Cost Water and Sanitation Interventions. Human Development Report Office Occasional Paper. A collaboration between the Swiss Tropical Institute and the World Health Organization. 2006/33. 63p.

Hutton, G., Haller, L. and Bartram, J. 2007. Global Cost-Benefit Analysis of Water Supply and Sanitation Interventions. Journal of Water and Health: A Journal of the International Water Association (IWA), London. pp. 481-502.

ICIS. 2008. The Market. Fertilizer News and Analysis. 2008 Issue. Available at: http://www.fertilizerworks.com/html/market/TheMarket.pdf Accessed 8 August 2008. 5p

IFA. 2004. World Agriculture & Fertilizer Demand, Global Fertilizer Supply & Trade 2004 – 2005 International Fertilizer Industry Association 30th IFA Meeting Santiago, Chile 12/2004. Available at: www.fertilizer.org/ifacontent/download/8148/124196/version/1/file/2004_santiago_summary_report_.pdf Accessed 6 August 2008.

IWA. 2006. Sanitation 21: Simple Approaches to Complex Sanitation, A Draft Framework for Analysis. International Water Association. London. 38p.

Johansson, M. and Kvarnström, E. 2005. A Review of Sanitation Regulatory Frameworks. EcoSanRes Publication Series, Report 2005-1. 60p.

Kar, K. 2005. Practical Guide to Triggering Community-Led Total Sanitation (CLTS). A practical guide for use by frontline extension staff, based on experience of facilitating CLTS in at least eight different countries in South and South East Asia and in East Africa. Institute of Development Studies, UK. 14p.

Kar, K. and Chambers, R. 2008. Handbook on Community-led Total Sanitation. Institute of Development Studies at the Univ of Sussex and Plan UK. 90p.

Kvarnström, E. and af Petersens, E. 2004. Open Planning of Sanitation Systems. Report 2004-3. EcoSanRes Programme, Stockholm Environment Institute. Stockholm. 34p.

Lopez, A., Mathers, C., Ezzati, M., Jamison, D. and Murray, C. 2006. Global Burden of Disease and Risk Factor.World Bank and Oxford Univ Press. 475p.

Mara, D., Drangert, J-O., Nguyen Viet Anh, Tonderski, A., Gulyas, H. and Tonderski, K. 2007. Selection of Sustainable Sanitation Arrangements. Water Policy. IWA. 9: 305-318.

Marshall, S. J. 2004. Flagging Global Sanitation Target Threatens Other Millennium Development Goals. Bulletin of the World Health Organization. WHO, Switzerland. 82(3): 234-5.

Mayunbelo, K. M. K. 2006. Cost analysis for applying ecosan in peri-urban areas to achieve the MDGs – case study of Lusaka, Zambia. MSc Thesis. UNESCO-IHE Institute for Water Education, Delft, the Netherlands. Available at: http://www2.gtz.de/Dokumente/oe44/ecosan/en-cost-analysis-lusaka-2006.pdf Accessed 30 December 2007.

McConville, J., 2008. Assessing Sustainable Approaches to Sanitation Planning and Implementation in West Africa. Licentiate Thesis, Royal Institute of Technology, Stockholm, Sweden. TRITA-LWR LIC Thesis 2043.

Mehta, M., Fugelsnes, T. and Virjee, K. 2005. Financing the Millennium Development Goals for Water and Sanitation: What will it take? International Journal of Water Resource Development. 21(2): 239-252.

Miguel, E. and Kremer, M., 2004. Worms: Identifying Impacts on Education and Health in the Presence of Treatment Externalities. Econometrica 72(1): 159-217.

Moe, C. and Rheingans, R. 2006. Global Challenges in Water, Sanitation and Health. Journal of Water and Health. IWA Publishing. London. pp. 41-57.

Morgan, P. 2007. Toilets That Make Compost: Low-cost, sanitary toilets that produce valuable compost for crops in an African context. Stockholm Environment Institute. 99p.

Mugabi, J., Kayaga, S. and Njiru, C. 2007. Strategic Planning for Water Utilities in Developing Countries. Utilities Policy 15: 1-8.

Nadkarni, M. 2002. Drowning in Human Excreta. Down to Earth, Centre for Science and Environment, New Delhi. Available from http://www.downtoearth.org.in/cover.asp?foldername=20020228&filename=Anal&sid=3&page=4&sec_id=7&p=1 Accessed 12 August 2008.

NETSSAF. 2008. Deliverable 36 http://www.netssaf.net

Nokes, C. and Bundy, D. 1993. Compliance and Absenteeism in School Children: Implications for Helminth

Control. Transactions of the Royal Society of Tropical Medicine and Hygiene 87: 148-152.

OECD. 2006. The Challenge of Capacity Development, Working Towards Good Practice. Organization for Economic Co-operation and Development, Paris. Available at: http://www.oecd.org/dataoecd/4/36/36326495.pdf Accessed 10 September 2008.

OECD. 2007. Unsafe Water, Sanitation and Hygiene: Associated Health Impacts and the Costs and Benefits of Policy Interventions at the Global Level. Working Party on National Environmental Policies. Organisation for Economic Co-operation and Development. ENV/EPOC/WPNEP(2007)8/FINAL. 35p.

Olbrisch, S. 2006. Optimierung der Abwasserentsorgung aus Siedlungsgebieten im subsaharischen Afrika: Ein Beitrag zur Stadtentwicklungsplannung. Diploma thesis. Rhenish Friedrich-Wilhelm University, Bonn, Germany.

Onumah, G.E., Davis, J.R., Kleih, U. and Proctor, F.J. 2007. Empowering Smallholder Farmers in Markets: Changing Agricultural Marketing Systems and Innovative Responses by Producers' Economic Organizations. Brief of ESFIM Working Paper 2. 5p. Available at: http://www.esfim.org/download/Brief_ESFIM_WP_2_english.pdf Accessed 6 September 2008.

Örtengren, K. 2004. A summary of the Theory Behind the LFA Method- The Logical Framework Approach. Sida. Stockholm Available from http://www.sida.se/publications Accessed 11 August 2008.

Outlaw, T., Jenkins, M. and Scott, B. 2007. Opportunities for Sanitation Marketing in Uganda. USAID, Hygiene Improvement Project (HIP). United States Agency for International Development. 62p

Prüss-Üstün, A., Kay, D., Fewtrell, L, and Bartram, J. 2002. Estimating the Burden of Disease from Water, Sanitation, and Hygiene at a Global Level. Environmental Health Perspectives 10(5): 537–542

Prüss-Üstün, A., Kay, D., Fewtrell, L. and Bartram, J. 2004. Unsafe Water, Sanitation and Hygiene. In: Ezzati M, Lopez AD, Rodgers A, Murray CJL eds. Comparative Quantification of Health Risks. World Health Organization, Geneva. pp. 1321-1352.

Prüss-Üstün, A. and Corvalán, C. 2006. Preventing Disease Through Healthy Environments: Towards an Estimate of the Environmental Burden of Disease. World Health Organisation (WHO), Geneva. 106p.

Prüss-Üstün, A., Bos, R., Gore, F., Bartram, J. 2008. Safer Water, Better Health : Costs, Benefits and Sustainability of Interventions to Protect and Promote Health. World Health Organization (WHO), Geneva. 54p.

Ridderstolpe, P. 2000. Comparing Consequences Analysis – A Practical Method to Find the Right Solution for Wastewater Treatment. Available from http://www.iees.ch/EcoEng001/EcoEng001_R4.html Accessed 11 August 2008.

Rosemarin, A. 2004. In A Fix: The Precarious Geopolitics of Phosphorus. Down to Earth (CSE). June 30, 2004. pp.27-31.

Rosemarin, A. and Caldwell, I. 2007. The Precarious Global Geopolitics of Phosphorus. Int'l Conf on Sustainable Sanitation. Dongsheng, China. Aug 27-31, 2007. Available at: http://www.ecosanres.org/icss/proceedings/presentations/76--RosemarinAugust252007.pdf Accessed 6 August 2008.

Rosemarin, A. 2007. The Global Sanitation Crisis - Why Should 5000 Dead Children per Day be a Taboo Subject? Loft Bookazine. December 2007. pp. 280-284.

Satterthwaite, D. 2003. The Millennium Development Goals and Urban Poverty Reduction: Great Expectations and Nonsense Statistics. Environments and Urbanization, 15/2, London. pp. 181-190.

Satterthwaite, D. and McGranahan, G. 2007. Providing Clean Water and Sanitation. In The State of the World: Our Urban Future. Chapter 2, Worldwatch, 24th Edition. Earthscan, London. pp. 26-45.

Schmid Neset, T-S., Lohm, U., Bader, H-P., Scheidegger, R. 2008. The flow of phosphorus in food production and consumption – Linköping, Sweden, 1870-2000. Science of the Total Environment 396:111-120.

Scott, C.A., Faruqui, N.I. and Raschid-Sally, L. 2004. Wastewater Use in Irrigated Agriculture: Coordinating the Livelihood and Environmental Realities. CABI/IWMI/IDRC. 206p. Available at http://www.idrc.ca/en/ev-31595-201-1-DO_TOPIC.html Accessed 14 September 2008.

SEI. 2005. Sustainable Pathways to Attain the Millenium Development Goals: Assessing the Key Role of Water, Energy and Sanitation. Stockholm Environment Institute (SEI). 103p.

Shah, F. 2007. Local Financing for Rural and Peri-Urban Water Supply and Sanitation. Protos. 13p. Available at: http://www.protos.be/protosh2o/copy_of_Links-met-bestanden/LocalFinRurPeriUrbWATSAN.pdf Accessed 15 September 2008.

Shilton, A. & Walmsley, N. 2005. Introduction to Pond Treatment Technology. In: Shilton, A. (ed). 2005. Pond Treatment Technology. IWA Publishing, London. 479p.

Shrestra, G., Tayler, K. and Scott, R. 2005. Assessing Nepal's National Sanitation Policy. Maximizing the Benefits from Water and Environmental Sanitation: 31st WEDC Conference, Kampala, Uganda. WEDC, Loughborough, UK. pp 76-79.

Steen, P. 1998. Phosphorus recovery in the 21st century. Management of a non-renewable resource. Phosphorus & Potassium Journal, Issue No. 217

Strauss, M. 2000. Human Waste (Excreta and Wastewater) Reuse. A contribution written for the ETC/SIDA Bibliography on Urban Agriculture. EAWAG. 31p. Available at: http://www.eawag.ch/organisation/abteilungen/sandec/publikationen/publications_wra/downloads_wra/human_waste_use_ETC_SIDA_UA.pdf Accessed 14 September 2008.

SuSanA. 2008. Towards more sustainable sanitation solutions. Version 1.2. Available at: http://www.susana.org/images/documents/02-vision/en-susana-vision-statement-I-version-1-2-feb-2008.pdf Accessed 8 September 2008.

Tannerfeldt, G. and Ljung, P. 2006. More Urban Less Poor. Earthscan. 192p.

Tayler, K. and Scott, R. 2005. Assessing National Sanitation Policy for Effectiveness: Lessons from Nepal and Ghana. Maximizing the Benefits from Water and Environmental Sanitation: 31st WEDC Conference, Kampala, Uganda. WEDC, Loughborough, UK. pp. 84-87.

Teka, G.E. and Metaku, M., 2003. Sanitation and Hygiene Issues Paper, 2 vols. Addis Ababa: WSP-AF, DFID and Ethiopian Ministry of Water Resources.

Tilley E., Lüthi C., Morel A., Zurbrugg C., Schertenleib R. 2008. Compendium of Sanitation Systems and Technologies. Eawag/Sandec & WSSCC, Dübendorf, Switzerland. Unpublished Preprint (in press).

Toubkiss J. 2006. Costing MDG target 10 on water supply and sanitation: Comparative analysis, obstacles and recommendations. 2006. World Water Council. Available at: http://www.worldwatercouncil.org/fileadmin/wwc/Library/Publications_and_reports/FicheMDG_UK_final.pdf Accessed 29 November 2007.

Ujang Z. and Henze M. (eds.) 2006. Municipal Wastewater Management in Developing Countries. Principles and Engineering, IWA, London. 352p.

UN. 2005. A gender perspective on water resources and sanitation. Background Paper, CSD. Available at: http://www.un.org/esa/sustdev/csd/csd13/documents/bground_2.pdf. Accessed 11 February 2008.

UN. 2006. The Millenium Development Goals Report 2006. United Nations, New York. 32p.

UN. 2007. The Millennium Development Goals Report 2007. United Nations, New York. 21p.

UN Millennium Project. 2005. Health, Dignity, and Development. What Will It Take? Earthscan & James and James, London, UK. 228p.

UNDP. 2006. Beyond Scarcity: Power, Poverty and the Global Water Crisis. Human Development Report, NY. 388p.

UNEP. 2002. Melbourne principles for sustainable cities. Integrated Management Series No. 1, UN Environmental Programme, Division of Technology, Industry & Economics, Osaka

UNEP. 2007. Global Environment Outlook. GEO-4. 540p. Available at http://www.unep.org/geo/geo4/media/ Accessed 6 August 2008.

UNEP, WHO, UN-Habitat and WSSCC. 2004. Guidelines on Municipal Wastewater Management. UNEP-GPA Coordination Office. The Hague. 92p.

UNESCO. 2006. Water a Shared Responsibility. Berghahn Books, New York. 584p.

UNESCO and GTZ. 2006. Capacity building for ecological sanitation: Concepts for ecologically sustainable sanitation in formal and continuing education. UNESCO Working Series SC-2006/WS/5. 156p.

UNFPA. 2007. State of the World Population: Unleashing the Potential of Urban Growth. United Nations Population Fund (UNFPA). New York. 99p.

UNICEF. 2006. Progress for Children: A Report Card on Water and Sanitation. Number 5, UNICEF, NY. 36p.

UNICEF. 2007. Progress for Children. A world fit foe children statistical review. Available at: http://www.unicef.org/progressforchildren/2007n6/index_41802.htm Accessed 15 January 2008

USGS. 2006a. Mineral Commodity Summaries 2006 Appendices. Available at: http://minerals.usgs.gov/minerals/pubs/mcs/2006/mcsapp06.pdf. Accessed 16 September 2008.

USGS. 2006b. Mineral commodity summaries: 2006 phosphate rock. Available at: http://minerals.usgs.gov/minerals/pubs/commodity/phosphate_rock/. Accessed 6 August 2008.

USGS, 2008. Mineral commodity summaries: 2008 phosphate rock. Available at: http://minerals.usgs.gov/minerals/pubs/commodity/phosphate_rock/. Accessed 6 August 2008.

Utzinger, J. and Keiser, J. 2004. Schistosomiasis and Soil-Transmitted Helminthiasis: Common Drugs for Treatment and Control. Expert Opinion on Pharmacotherapy 5(2): 263-285.

Valent, F., Little, D., Bertollini, R, Nemer, L., Barbone, F., Temburlini, G. 2004. Burden of Disease Attributable to Selected Environmental Factors and Injury Among Children and Adolescents in Europe. Lancet 363: 2032-2039.

von Braun, J. 2007. The World Food Situation New Driving Forces and Required Actions. Food Policy Report No. 18. IFPRI. 18p.

Von Sperling, M. & Chernicharo, C. A. L. 2005. Biological Wastewater Treatment in Warm Climate Regions. IWA, London, UK. 1496p.

Wagstaff, A. and Van Doorslaer, E. 2003. Catastrophe and Improvement in Paying for Health Care: With Applications to Vietnam 1993-1998. Health Economics 12 (11): 921-34.

Wagstaff, A. and Claeson, M. (eds.) 2004. The Millenium Development Goals for Health: Rising to the Challenges. World Bank, Washington, DC. 186p.

WaterAid. 2008a. Tackling the Silent Killer: The Case of Sanitation. London. 15p.

WaterAid. 2008b. Giving Sanitation the Green Light. London. 2p. Available at http://www.wateraid.org/documents/giving_sanitation_the_green_light.pdf Accessed 15 September 2008.

WaterAid Nepal. 2004. The Water and Sanitation Millennium Development Targets in Nepal: What Do They Mean? What Will They Cost? Can Nepal Meet Them?, WaterAid, Kathmandu, Nepal. 42p.

WEHAB. 2002. Water, Energy, Health, Agriculture and Biodiversity working group for the World Summit on Sustainable Development. A Framework for Action on Water and Sanitation, UN, NY. 40p

Werner, C. 2004. Ecosan – principles, urban applications & challenges. Presentation at the UN Commission on Sustainable Development, 12th session – New York, 14-30 April 2004.

Winpenny, J. 2003. Financing Water For All. Report of the World Panel on Financing Water Infrastructure. Chaired by M. Camdessus. World Water Council, 3rd World Water Forum and Global Water Partnership. 54p. Available at: http://www.worldwatercouncil.org/fileadmin/wwc/Library/Publications_and_reports/CamdessusSummary.pdf Accessed 15 September 2008.

Wiwe, S. 2005. Participatory Multi-Criteria Decision Making in Sewer Planning in Ecuador: Case Study Fanca (Bahia de Caraquez), Ecuador. Technical University of Denmark, Dept. of Environment & Resources: Masters Thesis Report, S973370.

WHO. 1997. Strengthening Interventions to Reduce Helminth Infections: An Entry Point for the Development of Health-Promoting Schools. WHO, Geneva. (WHO/SCHOOL/96.1 ed.). 29p

WHO. 2002. World Health Report. Reducing health risk, promoting healthy lives. Oxford University Press. 230p.

WHO. 2004. Study of Environmental Burden of Disease in Children: Key Findings. WHO Regional Office for Europe, Copenhagen. 7p.

WHO. 2007. A Safe Future. Global Public Health Security in the 21st century. World Health Report, WHO, Geneva. 96p.

WHO. 2008 – Sanitation related diseases. Available at: http://www.who.int/water_sanitation_health/diseases/diarrhoea/en/. Accessed 12 February 2008.

WHO. 2008. Almost a quarter of all disease caused by environmental exposure. Available at: http://www.who.int/mediacentre/news/releases/2006/pr32/en/index.html. Accessed 20 February 2008.

WHO & UNICEF. 2000. Global Water Supply and Sanitation Assessment Report. WHO and UNICEF, Geneva, Switzerland/New York, USA. 87p.

WHO & UNICEF. 2004. Meeting the MDG Drinking Water and Sanitation Target: A Mid Term Assessment of Progress. WHO and UNICEF Joint Monitoring Programme (JMP). 33p.

WHO & UNICEF. 2005. Water For Life: Making It happen. WHO and UNICEF Joint Monitoring Programme (JMP) for Water Supply and Sanitation. 44p.

WHO & UNICEF. 2006. Meeting the MDG Drinking Water and Sanitation Target: The Urban and Rural Challenge of the Decade. WHO & UNICEF Joint Monitoring Programme (JMP) World Health Organisation, Switzerland. 47p.

WHO & UNICEF. 2008a. Progress on Drinking Water and Sanitation: Special Focus on Sanitation. Joint Monitoring Report. 56 p.

WHO & UNICEF. 2008b. A snapshot of sanitation in Africa. A special tabulation for AfricaSan based on preliminary data from the WHO/UNICEF JMP for water and sanitation. 14p.

Winpenny, J. 2003. Financing water for all. Report for the World Panel on Financing Water Infrastructure. Chaired by M. Camdessus. Available at: http://www.worldwatercouncil.org Accessed 4 February 2008.

Wood, S., Sawyer, R. & Simpson-Hébert, M. 1998. PHAST step-by-step guide: a participatory approach for the control of diarrhoeal disease. World Health Organization. Geneva (document WHO/EOS/98.3).

World Bank. 2006. Factors Affecting Supply of Fertilizer in sub-Saharan Africa. Agriculture and Rural Development Discussion Paper 24. 66p.

World Bank. 2007. A visual guide to the world's greatest challenges. Atlas of Global Development. Collins. The World Bank, Washington, DC. 144p.

World Water Week. 2008. Session on "Monitoring Drinking Water Supply and Sanitation: Moving Beyond 2015, Preparing the Next Generation of Indicators". Available at: http://www.worldwaterweek.org/programme/thursday/thu18-movingbeyond2015.asp Accessed 14 September 2008.

Wright, A. M. 1997. Toward a Strategic Sanitation Approach: Improving the Sustainability of Urban Sanitation in Developing Countries. UNDP-World Bank Water and Sanitation Program. Washington. 38p.

WS Atkins International Ltd, Abt Associated Inc., CARE International and London School of Hygiene and Tropical Medicine. 2003. World Bank 'State of the Art' Hygiene and Sanitation Promotion Component Design for Large Scale Rural Water and Sanitation Projects. Inception Report.

WSP. 2007a. From Burden to Communal Responsibility. A Sanitation Success Story from Southern Region in Ethiopia. Water and Sanitation Program Field Note. Sanitation and Hygiene Series. Nairobi. 11p.

WSP. 2007b. Lessons from a Low-Cost Ecological Approach to Sanitation in Malawi. Water and Sanitation Program Field Note, Nairobi. 11p.

WSSD, 2002. Plan of Implementation 2002. World Summit on Sustainable Development in Johannesburg. Available at: http://www.un.org/esa/sustdev/documents/WSSD_POI_PD/English/WSSD_PlanImpl.pdf Accessed 6 September 2008.

WWI. 2005. World Watch Report - State of the World 2005. Redefining Global Security. World Watch Institute, World Watch Books. 237p.